Programming Language Cultures

Programming Language Cultures

Automating Automation

BRIAN LENNON

STANFORD UNIVERSITY PRESS
Stanford, California

Stanford University Press
Stanford, California

Printed in the United States of America on acid-free, archival-quality paper

Library of Congress Cataloging-in-Publication Data
Names: Lennon, Brian, 1971– author.
Title: Programming language cultures : automating automation / Brian Lennon.
Description: Stanford, California : Stanford University Press, 2024. | Includes bibliographical references and index.
Identifiers: LCCN 2024003848 (print) | LCCN 2024003849 (ebook) | ISBN 9781503633353 (cloth) | ISBN 9781503639874 (paperback) | ISBN 9781503639881 (ebook)
Subjects: LCSH: Programming languages (Electronic computers)—Social aspects. | Programming languages (Electronic computers)—History. | Language and culture.
Classification: LCC QA76.7 (print) | LCC QA76.7 (ebook) | DDC 005.1309—dc23/eng/20240220
LC record available at https://lccn.loc.gov/2024003848
LC ebook record available at https://lccn.loc.gov/2024003849

Cover design: Aufuldish & Warinner
Cover art: Bob Aufuldish

To David Golumbia
This book is dedicated to the memory of David Golumbia,
a polymath and a person of conscience

CONTENTS

Programming Language Cultures

INTRODUCTION

Eating the World

I

"More and more major businesses and industries," Marc Andreessen observed in "Why Software Is Eating the World," published in the *Wall Street Journal* in 2011, three years into the Great Recession, "are being run on software and delivered as online services—from movies to agriculture to national defense. Many of the winners are Silicon Valley–style entrepreneurial technology companies that are invading and overturning established industry structures." Andreessen, best known as the co-founder of Netscape Communications Corporation and the venture capital firm Andreessen Horowitz, used the phrase "software is eating the world" to describe the unexpected, yet undeniable explosion of so-called Web 2.0 or social media platforms amid the widespread economic suffering and the self-destructive austerity politics of the time, from which the tech industry appeared, this time, to have engineered a successful escape. "More than 10 years after the peak of the 1990s dot-com bubble," as Andreessen put it, "a dozen or so new Internet companies like Facebook and Twitter are sparking controversy in Silicon Valley, due to their rapidly growing private market valuations, and even the occasional successful IPO. With scars from the heyday of Webvan and Pets.com still fresh in the investor psyche, people are asking, 'Isn't this just a dangerous new bubble?'" "I, along with others," Andreessen told the *Wall Street Journal*'s readers, "have been arguing the other side of the case" (Andreessen 2011).[1]

It is accurate enough and unhyperbolic to state that since the early 2000s, a meaningful intensification—which is technical but also cultural, economic, and political in character all at the same time—has generalized an awareness of radical and profound infrastructural and organizational dependence on software. That is to say that possibly for the first time in a longer history of computing, such awareness has become very widely distributed outside the tightly coupled classes of technical experts we call software engineers and their managers. Now, that awareness is also alive among many of those who may not build software applications for a living, or directly manage those who do, but whose livelihoods in financial services, media, education, and other domains are equally inconceivable without the mediation of the software industry.

Only six years after Andreessen's pronouncement, things looked somewhat different, as the magnitude of the side effects of a narrowly software-driven recovery first came into view. "For the last year," Biz Carson wrote in *Business Insider* on November 12, 2016, "the tech industry has been fretting about a bubble. Investors on all sides argued over whether valuations were too high or whether the tech sector as a whole was still undervalued. Yet while Silicon Valley was obsessing over the startup bubble, it collectively failed to realize it was living in a completely different kind of bubble: a political bubble. As the reality struck late Tuesday night that Donald Trump would be the next US president, tech leaders found themselves reeling." Trump's election win, Carson concluded, deliberately echoing Andreessen's phrase, "signaled to Silicon Valley that it's time to look outside the bubble. Software has eaten the world, and this is what's been vomited back up" (Carson 2016).

Whether its motive is boosterish or critical, it is still common today to hear that software runs the world, that life as we know it is impossible without software, or that software is "eating the world"—not necessarily as a good thing, but certainly as an inevitability. Such statements aren't simple fictions. But they are often what I'd call marginally dishonorable, in two ways. Their first problem is that much of life as we know it has been running on software for nearly seven decades already. What we now call stored-program electronic computing, in which instructions for computation are represented as data in memory, dates to 1949. The UNIVAC I, the first successfully marketed commercial stored-program computer, was in wide use by 1954, installed at facilities operated by manufacturing giants including General Electric, U.S. Steel, Du Pont, and Westinghouse; insurance companies including Metropolitan Life, Franklin Life, and Pacific Mutual Life; utility companies like Consolidated Edison; US government agencies including the Census Bureau and Atomic Energy Commission; and divisions of the US armed forces including the Air Force, Army, and Navy. We tend to think of the 1960s as the decade in which the US economy was "digitized," but as James W. Cortada reminds us in his three-volume historical study *The Digital Hand,* the immediate and widespread commercial success of the IBM 650, introduced in 1953, suggests that the operations of big companies in the US manufacturing, transportation, retail, financial services, telecommunications, media and entertainment, and public sector industries were in many ways already reliant on computing by the end of the 1950s.[2]

That is to say that the long march toward the estimated 220 billion lines of mission-critical legacy code in the Cobol programming language still running on mainframe computers today, disproportionately at banks and other financial service providers, was well under way by 1959, when Cobol was first introduced.[3] There is, in other words, nothing especially new here. One does not hear similarly aggrandizing claims about nuclear energy, mass-produced plastics, color television, or other indispensable and ubiquitous technologies whose emergence also dates to the 1940s. Consider, for example, the brief introduction that Josh Tyrangiel, then editor of *Bloomberg Businessweek,* wrote for "What Is Code?," the article he commissioned Paul Ford to write in 2015 that became internet-famous as a 38,000-word interactive web feature comprising an entire issue of the magazine. "Software has been around since the 1940s," Tyrangiel wrote. "Which means that people have been faking their way through meetings about software, and the code that builds it, for generations. Now that software lives in our pockets, runs our cars and homes, and dominates our waking lives, ignorance is no longer acceptable. The world belongs to people who code. Those who don't understand will be left behind." While its rhetorical purpose is clear enough, none of the reasoning in this brief paragraph actually follows from the premise expressed by its first and second sentences.[4]

The second problem with a pronouncement like "software runs the world" is that it directly serves both the economic advantage and the generalized economic, political, and cultural authority of a very specific kind of technical expert, the computer programmer or software developer or engineer, whose economic role, along with those of other experts who directly support or exploit the programmer's work, has now been elevated beyond reasonable measure. It's safe to say that the violence of the companion phrase "eating the world," its assertion of destructive privilege, is an intended violence: one part opportunism, conscious or otherwise, and one part naked aggression, reflecting the succession of the Wall Street i-banker by the Silicon Valley tech bro in public economic-historical lore. Though their work conditions aren't perfect and they gripe like anyone else, software engineers went uniquely unscathed—indeed, made out like bandits—during the extended era of austerity and generalized economic pain and suffering that began in 2008. Many remained insensitive to the context of their good fortune, in some cases all the way up to the painful sobriety imposed by the unprecedented industry contraction that began in mid-2022. One ought not to join the chorus here, given how nakedly such talk reflects the extensive economic violence of that earlier interval, a period in which "learning to code" was imagined sometimes confusedly, but often cynically, as something like a universal pathway to reemployment.

II

While one cannot deny the importance of software, one might insist on the meaningful difference between the circular reasoning of such pronouncements—"software is important; important things deserve attention; therefore software deserves attention"—and the historical, economic, and political questions of *how* and *why* software came to be so important, along with the normative question of whether software's importance is, on balance, something good or something bad.[5] This book is a study of the powerful autologies or self-referentialities of programming language cultures, which programmatically elide such questions.

Autology is the opposite of *heterology*, a difference of origin or an incommensurability of parts. Computer programming is polyglot and translational, and no program, even a small one, can be written without resorting to various kinds and modes of bindings, hooks, and other established or improvised, approved or undocumented interfaces, forming a heterology both virtual and fragile. This is both true and obvious. I would suggest that these are not in fact programming's specially interesting characteristics, and they are not the characteristics it is urgent to document and analyze, today. Software engineers are in no way reluctant to admit, even to highlight, the persistence of bugs, crashes, security lapses and breaches, and all kinds of more serious disasters attending their work, and to acknowledge that at some level, given the technical dynamics of software production, such problems can never be completely eliminated. The practical activity of nearly every software engineer is nonetheless dedicated to avoiding or eliminating such disorder, because software engineering resources, incentives, and rewards are all aligned with that autological purpose. "Autology" here marks this continuous, insistent orientation toward purified self-reference, toward eliminating the disruption and alienation of error, which we find in the absolute centrality of debugging to programming: error, errancy, are always assumed, but *also* always opposed. The bug is hunted to be squashed, and the hunt never stops.

In a broader sense, *autology* refers to programming's special and privileged role in automation. Donald E. Knuth put it directly and unambiguously, remarking that "computer science answers the question 'What can be automated?'" (Knuth 1996, 3). While the most proximate reference of this remark is the emergence of the discipline of computer science from automated theorem proving, an early application of computers, Knuth clearly intended it as a generalization. The so-called programming language hierarchy, in which one abstract schematic after another, from mnemonics for operation codes to high-level languages incorporating English-language (or other natural-language) words and phrases, facilitates the automation of activities formerly performed manually, should be understood as an emblem of the generally recursive automation of programming as a political-economic activity: an activity that has no purpose but to automate other labor activities, not excluding itself. Automation of the jobs worth having: that's what a programming language culture does.[6]

And you can't understand the culture without understanding its language. On this point I agree with humanities-based and sociologically oriented critics and scholars in the domains of platform studies, software studies, and code studies whose perspectives I might otherwise characterize as technically or technologically deterministic. Today our general intellectual culture is almost completely paralyzed by technical ignorance of computing. In chronicling the process of "learning to code in middle age," Andrew Smith is correct to argue that "by accident more than design, coders now comprise a Fifth Estate and as 21st-century citizens we need to be able to interrogate them as deeply as we interrogate politicians, marketers, the players of Wall Street and the media" (Smith 2018). During the widely celebrated testimony of Facebook CEO Mark Zuckerberg to the US Senate Committee on Commerce, Science, and Transportation on April 10–11, 2018, there was much snickering among techie journalists, academics, and others who felt themselves in the know. But the truth is that while Gen Xers and millennials who grew up with BBSs, IRC, listservs, and then Web 1.0 still know quite a bit more than your average US senator about Facebook, Twitter, and the rest of Web 2.0, most are closer to the senators' ignorance than they are to true technical knowledge, particularly of the creation and maintenance of software infrastructures. (There was visibly less snickering during the testimony of TikTok CEO Shou Zi Chew to the US House of Representatives Energy and Commerce Committee on March 23, 2023, which among other things suggests how significantly the ideological hegemony of the tech industry had faded during the intervening five years.)

In this book, I have focused on consolidating and generalizing the latter form and mode of technical knowledge, of the creation and maintenance of software infrastructures, from a humanities-based yet not humanities-enclosed perspective, as a preface to—but certainly not an overtaking or elision of—a truly effective critique to come. To date, attempts to generate a political-economic theory of digital labor have focused more on semiprecarious digital piecework and so-called "free" contingent labor than on the relatively or distinctly elite work of the programmer.[7] Sociologically oriented business histories have offered valuably detailed documentation of the histories of computer programming and computerized data processing as elite professions, but less close or extended study of programming languages and their usage infrastructures in themselves.[8] Political philosophy generally, even in the accelerationist mode that enjoins the left to "develop sociotechnical hegemony,"[9] still remains remote from technical realities. While the same cannot be said either of philosophical approaches to software as media, or of culturally oriented research in the politics of computation, such work prioritizes theoretical interpretation and its scope tends to exceeds software as such,[10] though in this area at least one sociologically oriented cultural studies scholar has documented specific cultures of software development in work that is laudable for being simultaneously technically detailed and socially focused.[11] While scholarship in the literary humanities has been receptive to such research in so-called software studies, it has not displayed a proportionate interest in the specifically social and cultural dimensions of the specifically linguistic history of computing, as one would expect it to do—and this is the case especially where individual programming languages and their development and usage cultures are concerned.[12] The broad exception is, of course, in the narrative historiography of computing and in science and technology studies more broadly. But even here, Mark Priestley is surely right to suggest that early, purely technical histories of programming languages have been followed by socially attentive histories of software as a general object and domain, leaving individual programming languages behind as objects of potentially equally both technically and socially focused study (Priestley 2010, 2).

Granting that no duplication of the early, narrow technical histories is necessary—they were meticulous, if unsurprisingly disproportionately anecdotal in character[13]—how can we describe the humanities research space separating an early historiography of programming languages that is as old as the Fortran, Lisp, Algol, and Cobol languages themselves (which originates, that is to say, in the late 1950s), and recent social histories of the software concept such as Martin Campbell-Kelly's *From Airline Reservations to Sonic the Hedgehog: A History of the Software Industry* (2004)? A clue is to be found, I suggest, in an essay by William Paulson titled "For a Cosmopolitical Philology: Lessons from Science Studies" (Paulson 2001), insofar as in that essay, published in 2001, Paulson suggested the value of bringing science and technology studies (STS) scholarship into contact with an older literary humanist tradition of philology: a tradition whose methodologies were globally comparative and multilingual, whose mode was the study of texts in multiple languages (which required intensive study of the languages themselves), and which was rooted in a specific Western intellectual-historical tradition, the tradition of secular or historical humanism. If we set aside this latter tradition (one that STS scholars would surely understand themselves as sharing with "philologists," that is, with language and literature scholars), philology's characteristic mode of focus, grounded as it is in the mandates of linguistic specificity, even incommensurability, cannot be described as a great strength or even necessarily a normal characteristic of scholarship in STS.

From a position close to Paulson's own, one might invite software studies and so-called critical code studies, as well as STS itself, to establish an as-yet imagined contact with philology. How might we imagine a philological study—that is, a minimally both technically and socially oriented, but specifically linguistic historiography—of a specific computer programming language, or a specific characteristic of a type or family of programming languages, or the social and historical context of a particular programming language feature and its place in a culture of software development? For an example of how such questions might be posed within the disciplinary context of the information sciences, we can consult recent work like Amy J. Ko's "What Is a Programming Language, Really?" "In computing," Ko remarks, "we usually take a technical view of programming languages (PL), defining them as formal means of specifying a computer behavior. This view shapes much of the research that we do on PL, determining the questions we ask about them, the improvements we make to them, and how we teach people to use them. But to many people, PL are not purely technical things, but *socio*-technical things" (Ko 2016, 32). Still, essays like Ko's are remarkably few and far between, in the domain of the technical sciences as much as in the social sciences and the humanities—and often, as in this particular case, perhaps unavoidably perfunctory. Regardless of how we choose to explain it, Ko's conclusion that "other agendas, particular those that probe the human, social, societal, and ethical dimensions of PL, are hardly explored at all" (Ko 2016, 33) is certainly warranted.[14]

III

Put more generally, this book's goal is to demonstrate one way of integrating humanistically based but not humanistically constrained study of both the histories and the technical characteristics of programming languages into the broader cultural history of technical expertise in computing established by the research of Jennifer S. Light, David Alan Grier, Janet Abbate, Nathan Ensmenger, Margot Lee Shetterly, Mar Hicks, and Charlton D. McIlwain, among other scholars.[15] I know this research well and I admire it highly. It is the closest thing to an existing point of departure for my own work in this book, and so it is cited, both individually and collectively, at some key moments. Nonetheless, its contact with philology, as I have discussed it here, is glancing at best and generally negligible, employing neither the word nor the concept. This body of work is in no way diminished by that fact. As the product of scholarly training and dispositions that are historiographic, but not specifically *linguistic-historiographic*, its orientation to our common topic is certainly proximate, yet also orthogonal to my own.

Another point of departure might be represented by the work of Ted Nelson, Alan Kay, Seymour Papert, Michel Resnick, and other trained or untrained computer scientist-engineers whose consilient devotion to broadly pedagogical, creative, expressive, and other personal or personalist applications of computing won them many friends in the arts and humanities, as well as some imitators. I have written elsewhere[16] of the fundamental literariness of Nelson's imagination of a file system capable of accommodating and co-producing the true dynamic of knowledge as a "disappearance and up-ending of categories and subjects" (Nelson 1965, 96). Nelson's treatment of computers as humanist "dream machines" and "literary machines" represents an aestheticizing radicalization of the philological approach that goes well beyond my own inclinations in this book. The work of Papert, Kay, and Resnick on the Logo, Squeak, Etoys, and Scratch programming languages and environments, among other projects, is widely regarded as a more successfully institutionalized bridge from computer science to the humanities—and with good reason, despite the troubled legacy of the One Laptop per Child project with which all three men were involved.[17] More recently, the Racket programming language, environment, and associated pedagogical tools and materials, including the textbook *How to Design Programs*, have proposed reforms to postsecondary instruction in computing that carry the motives of Papert's, Kay's, and Resnick's projects forward into an age of ubiquitous and generalized computing.[18]

Here, too, this is research that I admire, and with which I have long been familiar. If it is less proximate as a point of departure for my own work in this book, and thus goes unacknowledged beyond this introduction, that is for reasons beyond its equidistant remoteness from philology. First, with the possible exception of Nelson, who deserves a category of his own, none of these researchers goes as far as Donald E.

Knuth in bringing *literature,* specifically, to meet programming, specifically, as I have described this part of Knuth's work in chapter 2—without
also tipping into potentially facile conflations of code with language, or
knowledge of programming with literacy. The latter two category mistakes, unfortunately widespread in the contact zone between programming and other kinds of writing, are problematically aggressive on the
humanities' own side of the border, where attempts to gather technology
into philosophy, and thus buttress the latter's primacy, are a longstanding project. By contrast, Knuth's concept of "literate programming," as I
describe it in chapter 2, borrows terms and concepts from literature as a
discrete and distinct domain, without subsuming programming in literature *or* subsuming literature in programming. Knuth's actual practice of
literate programming, meanwhile, focused on writing single programs
to be reproduced in two clearly separate and meaningfully different, yet
epistemologically equalized forms: the virtual object destined for execution, and a typeset document intended to be read as an essay. That Knuth
manages this without losing sight of the distinguishing characteristic of
programming as *automation,* the bête noire of all humanisms—indeed,
while defining computer science itself, paradigmatically, as the study of
what can be automated—is still more remarkable, both as rhetoric and as
thinking. More recent projects in a similar spirit, including Cristina Videira Lopes's two editions of *Exercises in Programming Style* and Angus
Croll's lighthearted but not unserious *If Hemingway Wrote JavaScript*
(2014), can be understood as variations of Knuth's specific disposition of
literate programming as truly conjoining these two domains, rather than
merging one into the other.[19]

Second, and more basically: individual disciplines and major clusters
of disciplines exist for historical reasons. Which is not at all to call individual disciplines, major clusters of disciplines, or the concept or system
of disciplines itself outdated or obsolete. On the contrary, this is rather to
observe that their borders are not simply imaginary—merely willed, as it
were, and thus arbitrary, even when we creatively exceed them, as some
of the figures I mention here did from the engineering side of the border,
and as I hope to have done within philology understood as an enclave.

Finally, a word about Mark C. Marino's *Critical Code Studies* (2020),
a study whose specific orientation toward the literary humanities may
leave it seeming closest to my own. To begin with where we agree: I fully
share and heartily endorse Marino's conviction that "if code governs

so much of our lives, then to understand its operations is to get some sense of the systems that operate on us. If we leave the hieroglyphs to the hierarchs, then we are all subjects of unknown and unseen processes" (Marino 2020, 4). Leaving aside the likelihood that we will still be the subjects of those processes even if we see and know them, I agree with Marino that cultivation in this area can make a difference. I also see value in "critical code studies" as devoted, in Marino's description, to reading computer programs with primary attention to meanings "locked in their technosocial context" (Marino 2020, 4).

In other words, we agree on the urgency of both these things: a general ability to read program code, and a specific ability to read it in a context meaningfully exceeding the strictly technical and self-referential. I take Marino's side in some disagreements with other scholars, as well. For example, I share, albeit much more strongly, Marino's relatively mild antipathy to the argument that program code *is* language: or paraphrased charitably, that writing in a programming language is similar enough to writing in a natural language that the promise of their imaginary affinities ought to outweigh their actual differences. Unfortunately for such excitable speculations, code is not language—full stop. And as Marino also argues, code is not poetry, either: or paraphrased charitably, there is nothing of particular value to be gained from excitably minimizing their differences or from conflating them outright. Most program code, Marino puts it correctly if perhaps too patiently, "is worlds away from literature, film and visual art" (Marino 2020, 49–50).

There are, however, significant differences in our approaches. On the one hand, Marino emphasizes, as I do, the learning of programming languages. On the other hand, Marino cites approvingly (Marino 2020, 46) Rita Raley's speculation, in an essay published eighteen years before Marino's book, about the prospect of programming languages meeting foreign language requirements in university humanities curricula.[20] Marino's approval of this idea is difficult to reconcile with his resistance to its implications elsewhere in the book, as noted above. Setting aside the question of whether Raley would take precisely the same position today, in the eighteen years since Raley's remarks first appeared in print, substitutions of programming languages for foreign language requirements have seldom gained traction, notwithstanding the encouragement provided by right-wing legislators and liberal centrist technocrats alike, with the backing of Silicon Valley edupreneurs. While part of the reason for the slow uptake here may be simple inertia, there are other reasons as well. Once again: code is not language, full stop.[21]

Let's say that an ambitious researcher in computer science, who cannot read German, finds that they need to learn German to read a small or historically obscure corpus of research in mathematics or engineering that promises to be useful to their work. By what logic could, or should, a researcher in the humanities learn the Java programming language for a supposedly analogous purpose? No one has ever written scholarship—or anything else, except programs—in Java. As a related question, should a serious researcher in computer science, or any other information technology–related field, be relieved of the requirement or the incentive to publish their research results in papers, reports, articles, or books—and simply publish undocumented programs instead? Most researchers in those fields would find this idea absurd. It makes perfect sense to require students to learn *both* foreign languages and programming languages, but not to substitute the latter for the former.

A more fundamental difference in our approaches is that while Marino emphasizes the learning of programming languages, it is ultimately a glancing emphasis, which Marino qualifies immediately by appeal to the unnecessarily embellished concept of "cyborg literacy"—meaning a state of affairs in which humanities researchers collaborate with other researchers who know programming languages, rather than learning programming languages themselves (Marino 2020, 46). Because research in the humanities is never urgent, I see no benefit in such arrangements. They might be fun, sometimes, but they'll never be necessary, and there will always be time for the would-be cyborg to just do their homework instead. Overwhelmingly, *Critical Code Studies,* the book and the new research field that Marino imagines launching with it, aims to correct the preferential attention generally given to *software* over *code,* and while I have my sympathies with this project, in the end what Marino calls *code* is an abstraction. As Marino uses it, but without acknowledging this issue consistently, the word *code,* properly speaking, always refers to text written in a particular programming language—not any programming language at all, or all programming languages at the same time. And the concept corresponding to the word "code" abstracts the specificity of that particular programming language, whatever it may be, much as the word and concept "language" itself, as we normally use it, abstracts the specificity of English, the language of publication of this book, or of any other natural language.

In other words, for Marino the rightful and corrective focus should be on code. For me, it is on programming languages. That, more than anywhere else, is where we diverge. As I see it, Marino separates the idea of "reading code" from the learning of programming languages in order to present code as another kind of readable text, in kinship with other kinds of readable text. This is a rather exhausting and possibly exhausted habit in the literary humanities, particularly in the inflection lent it by the enormously influential work of the German literary critic Friedrich A. Kittler, who is cited abundantly in *Critical Code Studies.* Of Kittler's obsessive, career-long deposing of literary language, his insistence that code can, may, does in fact compete with literary language, specifically, and will someday usurp it, it must be said: though its pretense is to almost precisely the opposite, this is, knowingly or otherwise, more than anything else an attempt to enclose and enfold code in the study of literature—to ensure the continuity of the study of literature first of all. In its intellectual debts and its bibliographic world view, Marino's book is in this way narrowly grounded, indeed cemented in a single, media-oriented corner of the literary humanities, despite its interdisciplinary ambitions.

My return to philology, as a reset or re–starting point for the study of programming languages, is not a recuperation of programming languages to the humanities generally, let alone to the study of literature specifically. To do that is to leave oneself unable to acknowledge the most fundamental characteristic and purpose of programming languages in their own context, the context in which they were originally developed and in which they continue to be used today. Whereas for Marino, "code since its inception has been a means of communication" (Marino 2020, 17), for me programming languages always were and are first of all systems of automation. Finally, in explicitly distancing "critical code studies" from Donald E. Knuth's concept of literate programming, Marino dramatically underreads and underestimates Knuth's work on that topic, reducing it to a discourse about the "readability" of code, which it certainly is not (Marino 2020, 41). In sum: Marino rejects the conflation of language and code, while also tolerating it. I simply reject it. Marino focuses mainly on code, and views code as communication, while I focus

primarily on programming languages, and view them as systems for automation. Marino positions code in the narrow purview of the twentieth-
century literary humanities, while I look back to the origins of philology
for a broader perspective, one that includes the humanities but isn't enclosed by them.

IV

My approach is broadly philological in three distinct senses of that
term, none of them reducible to its quotidian associations with either
historical linguistics or literary historiography. First, it emphasizes the
histories of both natural and formal languages, including programming
languages, in their individual specificities, over their abstract formal or
structural characteristics—if not as more important than the latter, then
as equally interesting and deserving of attention. Second, it regards individual natural and formal languages as carriers and sometimes shapers
of (specific) cultural histories. Third, it aims to integrate knowledge
from different disciplines without rejecting the difference of disciplines,
imagining that difference a mere artifact of bureaucracy, or demanding
that that difference be discarded or overcome.

Historiography and philosophy have always dominated humanities-
based research on computing, but philology is neither philosophy nor historiography. And the history of this word, as Margaret Alexiou has noted,
is marked by "a steady erosion of its wider" meanings (Alexiou 1990, 56).
In ancient Greek, where it was distinguished from philosophy as the love
of wisdom, *philology* encompassed the love of argument, the love of reasoning, the love of learning, the love of learned conversation, and the
love of literature (Alexiou 1990, 56). In modern Greek up to the eighteenth
century, it was an equivalent of Latin *litteratura,* "inclusive of all kinds of
writing (history, theology, philosophy, even the natural sciences)" (Alexiou 1990, 56). Its first recorded use in English, in the seventeenth century,
was similarly wide in scope, meaning love of learning in general, love
of literature and rhetoric generally and literature in particular (Alexiou
1990, 56). Later English usage significantly attenuated its range of meaning, narrowing it to historical linguistics or a "science of language," while
in German its scope expanded again, acquiring an association with so-
called writing systems: assemblages of the alphabetic, syllabaric, or logo-
graphic modes, forms, and appurtenances of spoken languages.

Even in this context, the consilience of German philology has been understood as consonant with a scholarly practice that is even older than ancient Greece. "It is generally assumed," Michael Holquist has written, "that the first writing system is found in cuneiform tablets unearthed in the city of Uruk 11 dating roughly from 3200 BCE. Sumerian, the language represented in these tablets, died out as a spoken language early in the second millennium. [. . .] The priests and scholars who kept the wisdom of the Sumerian past alive in the second millennium BCE are the first philologists" (Holquist 2011, 269–70). This formulation, which describes philology as the study of writing systems as mediating affordances for the transmission of knowledge, suggests that philology is, paradoxically yet not nonsensically, both broader and more focused than historiography. It stands not only for the historiography of writing systems, specifically, but for the study of writing systems' mediations of knowledge, including their mediations of the reflective temporal dimension of experience we call history and its formal study as an object by the practice of historiography. Historiography, the writing of history, can only be performed using a writing system—in a natural or human language. Philology includes the study of the writing systems used to perform historiography, as historiography itself does not.

Before it is anything else, a programming language is also a writing system, albeit for a formal language rather than a natural one, and software is written text before it can be anything else. It follows that a historiography of programming languages uses one writing system to describe another, but without reflection on that bipartition or bicamerality itself. That reflection, I suggest, is the task of a philology of programming languages.

So, philology is not historiography. Neither is it philosophy, which separates itself from historiography more completely. In its broadest meaning, philology already includes philosophy, as it includes historiography. Indeed, philology stands comprehensively for a general cultural secularization within which historiography emerges as a rival of myth, while philosophy challenges theology and religious authority. Too, it stands against what Pieter Verburg, writing of the nineteenth-century German comparative linguist Franz Bopp, called "the direct interference and arrogance of philosophy in matters of language" (Verburg 1998, 461), or philosophy's abstraction from the writing systems used to perform philosophy. As Sheldon Pollock put it:

Philology is [. . .] the discipline of making sense of texts. It is not the theory of language—that's linguistics—or the theory of meaning or truth—that's philosophy—but the theory of textuality as well as the history of textualized meaning. If philosophy is thought critically reflecting upon itself, as Kant put it, then philology may be seen as the critical self-reflection of language. (Pollock 2009, 934)

In the philosophy of media, the philosophy of technology, and the philosophy of software, philology resists the philosophical obfuscation that subordinates technical knowledge to its enclosure or enfolding in something else. In the influential work of the French philosopher Gilbert Simondon (1924–89), this tendency, designed to ensure the primacy or the supremacy of philosophy, is itself an example of "culture [. . .] as a defense system against technics," the object of Simondon's own critique (Simondon 2017, 15). Simondon's grandstanding denunciations of an ignorant or resentful "facile humanism" that fundamentally misunderstands machines (Simondon 2017, 15–16) served as the rhetorical furniture of a crypto-humanism that isolated and elevated philosophy as an envelope for technics:

It is only at the level of both the most primitive and the most elaborate of all thoughts, philosophical thought, that a truly *neutral* or *balanced* because *complete,* mediation between opposing phases can intervene. It is thus *philosophical thought* alone that can assume the knowledge, valorization and completion of the phase of technicity within the entirety [*ensemble*] of man's modes of being in the world. (Simondon 2017, xvii, emphases in original, translator's interpolation in original)

To be clear, Simondon was entirely correct to identify technical ignorance of computing as a serious and urgent practical, intellectual, and sociopolitical problem. I second Simondon's imagination of an effective "cultural reform" in general education and knowledge of technics, as a prelude to effectively regulating technics (Simondon 2017, 19, 21). However, I do not share or condone its motivation. It should not be part of a project to assure the primacy or the supremacy of philosophy. This is a powerful tendency and an entertaining pastime in the literary humanities, now as it was then. But it is not an effective one. Philology is a distinct mode for the analysis of the textual dimensions of computing as a research object in itself, and of programming languages in particular—as contrasted, for example, with computing's mere use as a research tool. In this role I believe it is more effective than philosophy.

Lest this be mistaken for a so-called postcritical approach to the topic, let me be clear that my point of reference and point of departure, here, is the revival of philology as counter-Enlightenment by literary humanists such as Edward W. Said, Paul Bové, and Aamir Mufti, among others, whose scholarship has always combined and balanced erudition with critique.[22] Nothing could be further than "postcritique" from my intellectual affiliations in general or my goals in this book, and while unlike my previous work, this book contains almost no polemic, the framings and endpoints of the stories I tell in each chapter make their critical commitments clear. For me personally, the relevance of the work of Bruno Latour, for example, is disqualified by the fantastically delusional and opportunistic essay "Why Has Critique Run Out of Steam? From Matters of Fact to Matters of Concern" (Latour 2004), and the same applies to the American anticritical movement that rode Latour's coattails for the two decades marked by the invasions of Afghanistan and Iraq, the reorganization of the US economy around new platform monopolies, the saturation of US culture by the priorities of libertarian Silicon Valley, and the far-right authoritarian populism that emerged from a toxic stew of resurgent jingoism, unreflective libertarian and centrist liberal technocracy, and the algorithmic operationalization of likeness and reciprocal amplification as principles, indeed conditions of social relations. Preferring it to Latour's superficially similar, but to me at least, politically incoherent ideas, I have informally, nonprogrammatically, and nonpedantically adopted Simondon's imagination of *technicity in ensembles* in a loose, secular mode divorced from the philosophical obfuscation that it carries in Simondon's work.

In its focus on language, philology describes the way that power is wielded *through* and *in* language. While programming languages are codes, not languages, perhaps we wish to misconstrue them as languages because they are sites and applications of power in similar ways. Software is created using programming languages, full stop. There is no other way to do it. And both the designs and the usage cultures of programming languages reflect the often enormously privileged positions of those who created them and who use them to earn outsized, sometimes outrageous wages. At the level of designed characteristics, a programming language that optimizes for speed, efficiency, and type safety over other qualities, like approachability, convenience, and ease of use, may reflect the priorities of a specific context of privilege, such as financial, scientific, and military applications of high-performance computing. At the level of designed applications, programming languages designed for data analysis and so-called machine learning, rather than for the building of websites and small applications or for creative and expressive applications of computing, may find their niches in technological infrastructures that automate extractive and punitive surveillance or actuarial, judicial, and carceral deliberation. And so on. Specific programming languages (some more than others) serve as access keys to such domains. The autologies or technical self-references and self-referentialities of programming language cultures, which are their default states, abstract away such considerations, while a philology of programming languages restores them to view.

V

With varying degrees of explicitness, this book makes the following three arguments:

1. Technical ignorance of computing is a practical, intellectual, and sociopolitical problem.

It is a *practical* problem because it ensures dramatically greater exposure to harassment, abuse, and crime facilitated by insufficiently understood personal information security principles, including but not limited to routine, unavoidable quotidian procedures like choosing passwords and configuring a home network router.

It is an *intellectual* problem because it encourages even the highly educated, such as professional academic researchers, to journalistically reproduce technology industry hoopla and meme engineering. A general example is the catachrestic abuse of the word *digital* in such phrases as "digital scholarship" and "digital humanities."[23] A more specific example is the combined panic and glee that greeted OpenAI Inc.'s first offer of web-based public access to its ChatGPT product, in December 2022.[24] Intellectuals are persons temperamentally unable to refrain from commenting on any apparently important issue, and even today most of their remarks on computing do more harm than good. It is reasonable to suggest that the rollout of ChatGPT and an accompanying highly orchestrated publicity storm reflects tech industry panic in the face of declining material fortunes on top of declining hegemony, and the credulity with which the commentariat consumed and amplified this event would be fairly described as conditioned to a so-called Pavlovian degree. The issues of surveillance, data privacy, data bias, and the automation of labor remain areas of persistent intellectual weakness and uncritical thinking, even among those who proudly parade their new interests in them, and an emerging stampede, at the time of writing, toward "critical artificial intelligence studies" unwisely accepts the inaccurate, but entirely industry-friendly canard that the emergent or generative behavior of so-called AI technologies takes place in so-called black boxes, at least partly inscrutable even to their creators. Despite my antipathy to Simondon's crypto-humanism, as I have characterized it above, I acknowledge that Simondon's philosophy of technology is grounded in technical realities and technical details. This is better than nothing. We can distinguish such an approach from the Heideggerian tradition of the philosophy of technology, in which the supremacy of philosophy is a goal and a purposive distraction from technical realities, rather than a shortfall of the effort to understand. At the time of writing, OpenAI's ChatGPT and its siblings and competitors, including Microsoft's Bing and Google's Bard products, are still widely described as "AI technologies" even by those who should know better. Ontologically speaking, these products are not in any meaningful way artificial intelligences, or even, with slightly more technical accuracy, neural networks or large language models (LLMs). Rather, *they are software* operating on data—and software consists only of code in identifiably specific programming languages, full stop. Any real understanding builds from the bottom up here.

Finally, it is a *sociopolitical* problem. To explain the relative instability of the major forms of democratic political organization today in its appropriate context and scope—that is, world economic and social history—is a far larger undertaking than my undertaking here. Still, it is a mistake to construe the social manifestations of that instability that most of us have observed in our daily lives, in recent years, as the products of partisan or "polarizing" rhetoric in themselves. Rather, they are products of algorithmic operations performed on instances, iterations, and accumulations of speech of those types, which amplify their reach in several different ways. On an individual level, even a modest technical understanding of the recommender systems at the core of the vast software infrastructures of Facebook, Instagram, Twitter, YouTube, and other notably destructive social media platforms can help users of such platforms shape our usage so that they remain tolerable to use, if we wish to continue using them. At a systemic level, some meaningful technical understanding will be indispensable to effective regulation of such platforms by legislators and to encouraging and expanding such regulation through activism.

2. Automation is not a myth.

Writing in the 1950s, an era when French intellectual culture was both overtly and covertly fascinated by cybernetics, the French philosopher Gilbert Simondon (mentioned above) opined with supreme confidence that "pure automatism, excluding man and aping the living, is a myth that does not correspond to the highest level of possible technicity: there is no machine of all machines" (Simondon 2017, xvi). Simondon had contempt for "worshipers of the machine" who imagined "that by increasing and perfecting automatism one would manage to combine and interconnect all machines among themselves, in such a way as to constitute the machine of all machines" (Simondon 2017, 17). But one might say that his entirely sensible resistance to such excitable speculations left Simondon unprepared or unable to recognize the fundamental logic of computer programming as the automation of automation, which was being attentively and accurately described at the same moment by Grace Murray Hopper in the United States (see my extended discussion in chapter 1). Measured against Hopper's technically precise and sociologically insightful observations about the ongoing automation of programming itself, Simondon's characterization of "automatism" as "a rather low degree of technical perfection" (Simondon 2017, 17), and his insistence that "automatism, and its utilization in the form of industrial organization, which one calls *automation,* possesses an economic or social signification more than a technical one" (Simondon 2017, 17), seem mistaken at best and philosophically evasive or sophistic at worst. When James Martin observed in *Application Development without Programmers* that "the number of programmers *per computer* is shrinking so fast that most computers in the future will have to work at least in part without programmers" (Martin 1982, xv) lest "the entire American workforce [. . .] be needed to program its computers" (Martin 1982, 2), the apparently economistic reductio ad absurdum of these remarks was in truth a basic observation about the technical logic of software as automation. We have never needed, and we will never need, everyone to learn to program computers, if only because the purpose of programming computers is to relieve us of the tedium of that activity itself.

Indirectly and directly, in forming my arguments here, I have also drawn on three studies published in the last quarter-decade, two having been published very recently. These three works, located in economic historiography, economic sociology or political economy, and digital cultural studies, respectively, all represent direct engagements with a recent or longer history of both public and expert discourse about automation during the nineteenth and twentieth centuries. In *Inventing Ourselves Out of Jobs? America's Debate over Technological Unemployment, 1929–1981* (2002), Amy Sue Bix recapitulates "the history of technological unemployment talk in the United States" (Bix 2002, 8) with its antecedents in nineteenth-century debates on the other side of the Atlantic triggered by the mechanization of manufacturing. Aaron Benanav's *Automation and the Future of Work* (2022) critiques a resurgent or new "automation discourse" that emerged in the 2010s across the entire US party and extra-party political spectrum in response to the historically most recent and legible manifestations of "chronic labor underdemand" (Benanav 2020, x). Similarly, and with the same critical targets in view, Luke Munn's *Automation Is a Myth* (2022) declares that the notion of "full" automation is a fantasy and "a long-running fable about the future of work that needs to be reconsidered" (Munn 2022, 2).

Though the first of these works is a sober historiography and the second and third intellectual-political polemics, all three focus centrally on discourse, even when they support their rhetorical analyses, sometimes extensively, with reliable economic data. That often extensive support does not as often extend to *technical* realities, however, which means that all three risk imagining automation, at least for their readers, as *more* a discourse than anything else. But the fact is that software automation, specifically, is a specific technical reality, with a specific technical logic that deserves recognition. This does not mean that *Programming Language Cultures: Automating Automation* is not *also* fundamentally a cultural study, like the three works I have just mentioned here. However, as a study of programming languages as cultural artifacts, this book does give special attention to the specific technical logic of software as automation, a topic that each of those other three works largely elides, each for its own reasons.

In 2021, a year before OpenAI released the product it calls ChatGPT, best described as a consumer-friendly web interface and API (Application Programming Interface) to a software system that generates natural-language prose in response to natural-language prompts, the company released a similar product named Codex, designed to write code in widely used programming languages. While the public reception of Codex was considerably quieter than that of ChatGPT, its long-term impact can be expected to be far greater, given that the natural-language output of ChatGPT has a more limited range of truly plausible long-term and lasting real-world applications. Because the purpose of software is first of all automation, computer programmers have always worried, entirely reasonably, that they will invent themselves out of their jobs. While the adoption and integration of Codex and similar systems into the production of software may be unlikely to threaten the jobs of the most elite and highly skilled programmers, especially those who program these systems in the first place, if such systems prove persistent and durable there is every reason to expect them to impact job availability and wages for entry-level and moderately skilled programmers and those whose specializations are no longer in demand.[25] If one asks the ChatGPT system to assess the likelihood of such developments, one receives an answer carefully crafted to minimize controversy: for example, "In fact, one of the main goals of Codex is to augment and enhance the work of human programmers by automating repetitive or mundane tasks and providing intelligent suggestions and insights."[26] Because the work of human programmers already is, and always has been, the automation of repetitive and mundane tasks, with the definition of "repetitive and mundane task" constantly being shifted forward, this will be transparently cold comfort to any programmer who grasps the fundamentals of computing. The software-computational problem that still requires human ingenuity today is always becoming the repetitive and mundane task of tomorrow. That is how software development works. ChatGPT's answer on behalf of its sibling product is an excellent illustration of the logic of the automation of automation, or layered, even recursive automation.

3. Automation is a moving target.

In line with Munn's (2022) more considered, less rhetorically charged arguments that automation is always incomplete and unevenly distributed (which I discuss in chapter 6), an overt assumption of this book is that automation is a moving target rather than something that has been completely achieved. Three factors are pertinent here. First, general technological progress is ongoing, if slowly and steadily rather than dramatically, and this means that the scope of software automation, specifically, will continually expand even if it never reaches the "fullness" of completion or totalization. Second, automation requires constant maintenance. Software automation is buggy, in the first instance, and quickly becomes outdated, even with attentive maintenance. However, this does not obviate the technical logic of software as the automation of automation. Any insistence otherwise is likely to be driven by excessive rhetorical and ideological investment in the "resistance in the materials" being automated. What it *does* mean is that in line with broader refinements of technology in general, any software automation needs ongoing development to remain viable. Even a task that appears to have been fully automated still requires attention, in this sense. Third, automation has both immediate and longer term economic, social, and political impacts that influence the scope of its implementations and applications, which may advance, recede, or withdraw entirely in some cases, depending on circumstance. Increased productivity and efficiency is one side of the coin, while technological unemployment and social disruption is the other. Nothing is inevitable here.

VI

In Chapter 1, "A Third Language: Computing, Translation, Automation," I consider the translation metaphors employed by the first programmers of electronic computers as they both borrowed and invented terms for describing the writing and execution of programs. The phrase "higher level language," now widely used to describe the notation systems that emerged in the mid-1950s combining algebraic expressions with English-language keywords, refers to such systems' design for abstraction of hardware-dependent numeric and alphanumeric operation codes. Before such technical categorical terms as "higher level" had settled into common usage, early programmers borrowed liberally from the national and comparative philological domain of natural or human languages to describe relationships among activities and processes in computing and what would later come to be called computer science. Words like "foreign," "native," "translation," and "translatability" were widely used, during the 1950s, alongside or in combination with new terms from an emerging technical lexicon. Much of this borrowing is at odds with humanistic linguistic and literary translation theory insofar as the most influential forms of the latter have rejected analogies binding code to language. I suggest that their point of divergence is marked by the concept of automation. The history of computing in general and of computer programming in particular, I argue, is a history of generally recursive automation: that is, a history of the addition of successive layers of control through which higher and lower level codes are "translated" to each other, and which has as its vanishing point, both by design and by accident, a historical moment when the human knowledge of how to write them recedes. The anticipation of such a transition, its representation, and the formation of a discourse representing it, at a moment in the history of computing, is my topic in this chapter. That moment anticipated some problems of labor and economic organization we have struggled with much more recently, reflected in the growth of initiatives to add instruction in software programming to primary and secondary educational curriculums.

This book's second and third chapters concern contact zones within a computer program where natural languages and programming languages appear to touch or intermingle. In chapter 2, "Really Reading the Code, Really Reading the Comments," I continue the discussion of automation begun in chapter 1, arguing that computer programming can be understood both as a kind of writing and as a special class of human labor activity facilitating the automation of other human labor activities, including itself. From the beginning of the history of computer programming, I suggest, it was recognized that the technical logic of programming is a logic of automation, indeed a logic of self-automation or even recursive automation. In the body of the chapter I examine *program comments,* or free-form natural-language writing inserted into a program and intended only for human readers of that program, similar to the marginal and interlinear commentaries familiar to humanistic scholars of manuscript cultures preceding the generalization of the printing press (though not reducible to that similarity). I examine the ways that programmers use this space of relative freedom to document not only the code of a program but also the activity and the experience of programming as a job and a profession. In this way, I argue, program comments serve as a register of programming's fundamental logic of automation, leaving traces in code that are legible if one knows, or learns, where to look.

Chapter 3, "Etymologies of Foo," argues that the English-language nonsense words "foo," "bar," "baz," and others in a more or less standardized sequence of so-called metasyntactic variables commonly used in computer programming ought to be understood as meta-abstractive: abstracting the abstraction, as it were, of higher level programming languages and of the programming language hierarchy itself, leaving the self-automation of programming legible in a manner that rewards culturally oriented study. As a designed system, a programming language is determined by constraints incommensurable with the arbitrary historical finitude of a natural language. Only its reserved words or identifiers are lexed and parsed, or compiled as code to be run. Other text in a program represents data, is ignored (for example, program comments, as discussed in chapter 2), or violates rules of syntax, producing compiler error. Finally, there are identifiers chosen by the program writer as names for classes, functions or methods, and constants or variables. Provided that they are created correctly, according to specified (and rigid) orthographic rules, and that they are not identical to reserved words or identifiers, such names might be words in any natural language whose natural-language meaning, as presented by a dictionary, clarifies a program segment's intent, purpose, or structure. I argue that in a domain of such radical constraint, this freedom to choose names becomes a site of struggle, even a kind of suffering, as the programmer is faced with collaborating ad hoc a second, entirely separate, exclusively human-readable task- and program-specific semantics interlayered with that of strictly determined machine-readable syntax, in those portions of the program that run—and that this is why such putatively free naming is the object of so much legislation in programming language style guides, manuals of software craftspersonship, and management theory focused on cultures of software development.

This book's fourth and fifth chapters are studies of individual programming languages. Chapter 4, "Snobol: A Rememory of Programming Language History," is a study of the Snobol (StriNg Oriented symBOlic

Language) family of programming languages, widely used from the 1960s through the 1980s but almost entirely obsolete today. A so-called right-brain language (that is, one supposedly designed for inductive and creative applications) in an era when cultures of software development were understood to need managerial discipline above all else, Snobol was presented as relatively unconstrained and deliberately easy to use at a moment when the software crisis of the 1960s was pushing programmers and their managers in the opposite direction. When it is remembered at all, today, Snobol is often remembered as a language for experts whose specific expertise was *not* in computing and was not technical in the usual sense: that is, which lay in the nontechnical knowledge disciplines, especially the humanities. I argue that Snobol's orthogonality cannot in fact be satisfactorily explained by appeal to its imagination as a humanities programming language. Indeed, Snobol appears to have acquired both that domain of use and the reputation that went with it quite accidentally, which makes for an intrinsically instructive history in itself.

Chapter 5, "JavaScript Affogato: Negotiations of Expertise," is a study of the JavaScript programming language, which appeared shortly after the Snobol languages fell into obscurity and which developed into the most widely used programming language in the world today, a history that rather neatly inverts Snobol's evolution from a project of and for computing experts and specialists into a utility for humanities scholars. From its introduction, I argue, JavaScript's reception by software developers, and its importance in web development as we understand it today, was structured by a continuous negotiation of expertise. I use the term "improvised expertise" to describe both conditions for and effects of the unanticipated development of JavaScript, originally designed for casual and inexpert coders, into a complex of technical artifacts and practices whose range and complexity of use has today propelled it into domains previously dominated by other, often older and more prestigious languages. "Improvised expertise" also marks the conditions for and effects of specific developmental dynamics in JavaScript's recent history, which I review in detail in this chapter of the book, and which can be understood as having a metonymic relationship with the developmental dynamics of the software industry as a whole. As a programming language designed for accessibility to both novice programmers and nonprogrammers that outgrew its design unexpectedly and violently, developing into a language of high complexity that requires great skill to use effectively, JavaScript illustrates the typical and unsurprising refinement of technical expertise over time.

Chapter 6, "DevOps Fiction: Workforce Relations in Technology Industry Novels," serves as a coda and afterword, offering this book's only explicitly literary-critical exploration of its major themes. This chapter surveys the literary expression of the spontaneous management theory associated with the concept of "DevOps," or "development operations," a systematic focus on managing cultures of software development that emerged during the 2000s. I propose that we regard DevOps fiction as a legible category of contemporary literature, related to but meaningfully distinct from "Silicon Valley fiction" as its superset, and that rather than sensationalizing either the elite consumption of new digital, social, or platform products or their production by elite software engineers, DevOps fiction imagines the management of routine maintenance without which neither is possible, and which best expresses the routinization of programming's fundamental logic of automation. In its focus on the maintenance of the production of software and the internal "ops" through which cultures of software development are managed, I suggest, DevOps fiction is best understood as a literary branch of the line of software project management commentary that begins with Gerald M. Weinberg's *The Psychology of Computer Programming* (1971), a management handbook that Weinberg had originally intended to compose as a novel, and Frederick P. Brooks Jr.'s widely influential memoir *The Mythical Man-Month: Essays on Software Engineering* (1975).

FIGURE 1. "Semi-Automatic Digital Computation." Source: Charles W. Adams and Stanley Gill, *Digital Computers: Business Applications.* Reprinted with the permission of Massachusetts Institute of Technology.

ONE

A Third Language

COMPUTING, TRANSLATION, AUTOMATION

From Semi-Automatic to Automatic Digital Computation

The sketch in Figure 1 represents a transformation of the economic and cultural role of computing during the early 1950s. Published with the title "Semi-Automatic Digital Computation," it depicts Joe, a conspicuously harried middle-aged white male desk worker who, sleeves rolled up and both hair and necktie askew, perspires profusely in an amalgamation of cognitive effort, anxiety, and haste as he taps a clacking desktop mechanical calculator with the fingers of his left hand and writes in a notebook with a pen held by the fingers of his right. A stylized spiral of vertigo rises from the top of his head, while an eddy of arithmetic expressions orbits his temples. While one of of two disembodied hands removes a batch of paper from the front of Joe's desk, the owner of the hand demands Joe's response to his next assignment, problem number 277, in a "rush—P.D.Q.—in a hurry!" A second hand, meanwhile, places a fresh batch of paper on the desk, accompanied by the annotation "Here's problem #278, Joe!"

Onto this comic representation of a human computer are projected the now familiar design metaphors of a stored-program computer architecture. The sleeve of the second disembodied hand, the one delivering the fresh batch of paper, is labeled INPUT; that of the hand removing paper is labeled OUTPUT. The mechanical calculator on Joe's desk is labeled ARITHMETIC ELEMENT, the paper notebook STORAGE. The label CONTROL points to hapless Joe's head itself. (Behind him is a bookshelf labeled SECONDARY STORAGE, containing bound volumes with the titles NOTES, NUMERICAL METHODS, and LOGS.)

In the so-called digital age that we imagine we inhabit today, it is amusing to recall the semantic range of that now thoroughly reified English word, a range that draws back to the five physiological organs of manipulation and sensation attached to the human hand, if not still further.[1] Certainly, that range is the object of allusion, in this sketch, not least in the conspicuously splayed organs of Joe's left hand, the ones tapping the keys of the calculator. For Joe is not only digitally computing, here; he is himself also a computer, in the earlier, explicitly vocational sense of the word denoting a professional role or job. We are now familiar with the specifically gendered history of such roles and jobs: a history that this representation of this computer as a male desk worker might leave merely implicit, but which the next illustration I will discuss restores.[2]

The implication is, of course, certainly not that all or most business computation or other data processing was still done manually with desktop mechanical calculators in the early 1950s. Rather, we might best describe this sketch as an anthropomorphization of the design architecture of an electronic computer that represents a human activity as a problem, because it is human and thus limited by what is human. Alternately, we might say that it represents a burdensome or otherwise unwelcome human responsibility in a specific context.

The aforementioned sketch was one of a pair; Figure 2 provides its companion. It is a chart depicting the flow of communication in *automatic* digital computation—by implication, fully or nearly fully automatic digital computation, by contrast with the semi-automatic digital computation depicted in the first sketch. The chart emphasizes, indeed insists on the legibility of the labels marking components and their spatial or temporal relationships, and uses what extra pictorial space is available to provide explicit explanation (for example, of the function of the control element, which "takes instructions from storage then directs all other elements properly"). Big, sweaty Joe, the first sketch's unambiguous focus, has been replaced by two tiny stick figures, neither of them in any way

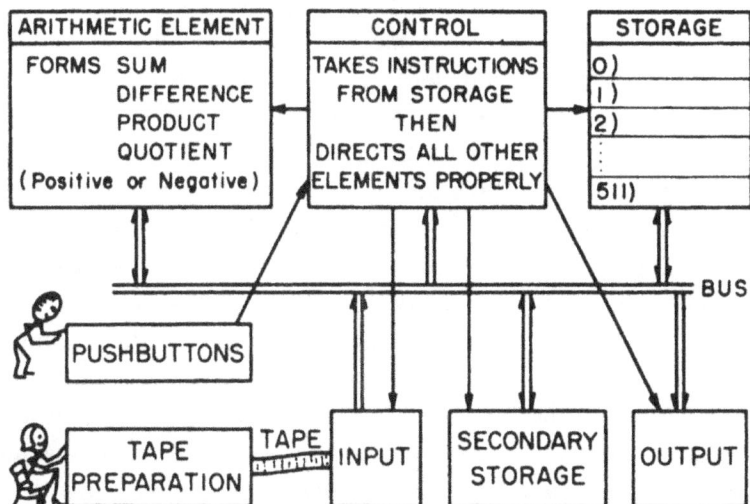

AUTOMATIC DIGITAL COMPUTATION

FIGURE 2. "Automatic Digital Computation." Source: Charles W. Adams and Stanley Gill, *Digital Computers: Business Applications.* Reprinted with the permission of Massachusetts Institute of Technology.

evidently anxious or unhappy, or encountering any mechanical or procedural difficulties. The stick-hands of the first figure, dressed in nothing but some short head hair, are connected to the label PUSHBUTTONS; those of the second figure, who wears a bob hairstyle, skirt and heels, and is seated in a chair, are affixed to the label TAPE PREPARATION. (The stick figure that we are expected to imagine as male is operating the machine's controls, while the figure we are expected to imagine as female is preparing data for it.) These two plainly less substantial and less consequential human figures are the deliberately puny, deliberately abstracted operators of the deliberately magnified electromechanical abstractions labeled INPUT, OUTPUT, CONTROL, ARITHMETIC ELEMENT, STORAGE, and so on.[3]

The two sketches were published together in "Digital Computers: Business Applications," a bound 263-page typescript of notes from a 1954 MIT summer session program with the same name. The notes accompanied one of a sequence of summer courses organized by Charles W. Adams at MIT, beginning with a 1953 programming course for which Adams devised the so-called Summer Session Computer, a hypothetical (that is, imaginary and simulated) computer for classroom instruction. Adams, who had worked on the Whirlwind project at MIT and would later found the Keydata Corporation, has been credited as having created the first "comprehensive conversion routine for translating coded orders into machine language" (Wildes and Lindgren 1985, 339): that is, for turning a human-readable so-called programming language into machine-executable operation codes or other instructions.[4]

In the pages preceding the page on which these two illustrations appeared, Adams, Stanley Gill, and their colleagues distinguished a digital computer from an analog computer in two distinct, conventional ways. First, they wrote, an analog computer measures continuous values, whereas a digital computer tabulates discrete values. Second, and in a sense tied more closely to behavior and purpose than to design, the behavior of an analog computer is analogous to a specific physical system. That is, an analog computer is a special-purpose device, whereas a digital computer has an at least potentially general purpose, which includes the ability to mimic a special-purpose device (Adams and Gill 1954, 1–1).

But it was not this distinction between analog and digital that really mattered. Certainly, a desk calculator such as the one operated by Joe, in the first illustration, might be understood as a special-purpose device in one context: that is, it would not be directly adjustable to provide, say, fire control for artillery, if such a function were desired. A desk calculator was analog in that second sense tied to behavior and purpose. But in another context, such as a context in which all meaningful input consisted of discrete values, a desk calculator could be understood as a general-purpose computer, and thus as digital in that second, looser sense of the word.[5]

The distinction important to Adams and his colleagues was not the distinction between analog and digital as much as it was the distinction between human and mechanical, or between manually performed and automated processes. "A competent person operating a modern desk calculator," they wrote,

> performs about 500 operations a day. By building a calculator a million times as fast, one can reduce the maximum of 5000 seconds of

machine operating time per day to a twentieth of a second, but speed up the overall process by at most 10% or 20%. The bottleneck is, of course, the human operator.

The automatically-sequenced digital computer is simply a mechanization not only of the arithmetic operations but of the operator who determines the sequence in which the operations are performed [. . .].

The arithmetic element of an automatically-sequenced digital computer, corresponding to the desk calculator of a manual system, can advantageously be made to work very fast, performing arithmetical operations in a few millionths of a second, for the rest of the system can now keep up with it. The control element, the counterpart of the human operator, can readily be made far faster, more reliable, and somewhat less demanding of wages and fringe benefits than the man. (Adams and Gill 1954, 1-5)

What Adams and his colleagues were interested in, and what these two cartoons (inserted at precisely this point in the typescript) were meant to depict, was the automation of a specific category of human activity: mechanically or electromechanically assisted but otherwise manual (that is, hand-performed) arithmetic calculation, and data processing more generally. That much, I trust, is obvious so far. But when the "bottleneck" of the human operator, as they imagined it—and as poor, big, sweaty Joe illustrates it, here—is backlit by the history of computer programming, a history that had only just begun, at this particular moment, it seems to shadow some problems of labor and economic organization we have struggled with much more recently. More than arithmetic calculation or other data processing as applications of a general-purpose digital computer, it was the *programming* of such a general-purpose digital computer, as preparation for such applications, that Adams and his colleagues imagined automating here. That is to say that both more specifically and more generally, as I shall emphasize in what follows, what Adams, Gill, and colleagues imagined, already in 1954, was the automation of the human activity of computer programming as we now know it today, and as we understand it as a special class of human labor activity facilitating the automation of other human labor activities, not excluding itself. It is the automation not only of computation but of computer programming itself, I am suggesting, that the shift from "semi-automatic digital computation" to "automatic digital computation" anticipates, in this pair of illustrations published in 1954. This represents an understanding of the intrinsic technical logic of programming that goes missing, for example, from the philosophico-technical speculations of the French philosopher Gilbert Simondon (discussed in the introduction), whose influential study *Du mode d'existence des objets techniques* (*On the Mode of Existence of Technical Objects* [1958]), which rejected automation as a "myth," was being composed at the very same moment. More properly, it is the anticipation of such a transition, its representation, and the formation of a discourse representing it, at this moment in the history of computing, that is my topic here. Understood as a preliminary automation of the work of the programmer, the compilation of a program written in a higher-level language may seem like a specialized application of software, but it is in fact software's core purpose and raison d'être, and it bears a direct relationship to the classical concept of automation understood as the reduction of human labor. Only the higher-level language in this case is written by the programmer, while the compiler performs automatic code generation. There is no intrinsic limit to the hierarchy of possible higher-level languages that might automate the writing of code at the next lower level. To be sure, a compiler is often a program written by a programmer, at least initially. But not always, and not necessarily: the historical co-emergence of compilers and so-called compiler generators or compiler-compilers, meaning compilers which themselves generate compilers, vividly underscores the intrinsic logic of software as the automation of the automation that is programming itself.

Programmer, Coder, Translator

My focus is on the translation metaphors and the translative metaphors employed by early computer programmers as they borrowed or invented words and phrases and established usage patterns in describing the writing and execution of programs. Programmers today are most likely to describe the virtues of any particular higher-level programming language in terms of its expressivity, that is, in terms of the range of activity afforded a fluent programmer in both describing and solving a computational problem using that language. The first and most important requirement of what was then called a "universal code," by contrast, was not expressivity but translatability—the facility, that is, with which its design permitted a programmer's written code to be automatically rewritten using the partly mutually incompatible operation codes of more than one hardware computer design and installation. The phrase "higher-level language," now widely used to describe the notation systems that emerged in the mid-1950s combining algebraic expressions with English-language keywords, refers to such systems' design for abstraction in this specific sense, rather than expression as such or more generally. The higher level they established hierarchically superseded a preceding abstraction, the hardware-dependent assembly language operation codes that represented binary digital instructions using alphanumeric mnemonics.

Before such technical categorical terms as "higher-level" had settled into common usage, programmers borrowed liberally from the national and comparative philological domain of natural or human languages to describe relationships among activities and processes in computing and what would later come to be called computer science. Words like *foreign*, *native*, *translation*, and *translatability* were quite widely used, during the 1950s, often alongside or in combination with new terms from the nascent technical lexicon now familiar to anyone with even a casual understanding of programming. A human computer programmer might, for example, be described as providing instructions in a "foreign language" that was made "translatable" for different computer installations by a special category of "translator" programs (later to be more consistently called "assemblers" or "compilers").[6]

The point where these domains of usage diverged is marked by the concept of automation. The "translation" performed by "translator" programs, from higher-level language to assembly or executable machine code, would come to be called "automatic coding," a much less ambiguous phrase denoting the automation of the human labor activity of formulating and transmitting lower-level hardware instructions.[7] "When a human operator is to solve a problem using a calculator or to process a payroll on an accounting machine," Adams and Gill wrote in *Digital Computers: Business Applications*:

> he must be supplied with instructions which specify just how the solution is to be obtained. In like manner, the digital computer must be provided with a list of instructions, in properly coded form, to describe how the solution is to be obtained. The process of preparing such a coded program is called programming. (Adams and Gill 1954, 1–7)

Adams and Gill used the word "programming" to refer to a two-step process in which a program is first planned, then written as "a coded program—a sequence of computer instructions" (Adams and Gill 1954). Though this distinction is elided in informal English usage today, which makes "coding" a less formal but straightforward equivalent for "programming," similar distinctions between these two words' ranges of meaning, as well as their potential conflation, were objects of explicit reflection through the 1950s.

In "Automatic Programming—Definitions," a paper delivered at the Symposium on Automatic Programming for Digital Computers convened the same year as Adams's MIT summer course on the business applications of digital computers, Grace Murray Hopper reflected on the decade that had passed since she first programmed the Harvard Mark I computer:

> In the ten years since Mark I first ran, the terms programmer and programming have come into being. The world is so accustomed to talk of automatic computers—large-scale, high-speed, automatically sequenced, digital, computing devices—that the adjective automatic has been dropped and only the term computer is required. Instead, the word automatic now is attached to programming and coding. (Grace Murray Hopper 1954, 1)

"Some nine or ten years ago," Hopper continued, "a programmer was, of necessity, many things. Numerical analyst, encoder-decoder, electrical or electronic technician, detective and bug-hunter, and finally, evaluator." With the design and implementation of more computers and the diversification and division of the labor of both their maintenance and their use for data processing, she observed, the roles of "analyst, programmer, coder, operator, and maintenance man" had gradually separated. Nonetheless, Hopper suggested, "the distinction between the programmer and the coder has never been clearly made. Coder was probably first used as an intermediate point on the salary scale between trainee and programmer" (Grace Murray Hopper 1954, 1).

A programmer, Hopper suggested, was someone who "prepares the plan for the solution of a problem," which may or may not include "numerical and systems analysis," but will invariably describe the way that the flow of data and operations should be managed. The programmer communicates such a plan to a coder in the form of a flow chart, and "it is then the task of the coder to reduce this flow chart to coding, to a list in computer code, the code representing the operations built into the hardware of the computer." "It is this function," she remarked, "that of the coder, time-consuming and fraught with mistakes, that is the first human operation to be replaced by the computer itself" (Grace Murray Hopper 1954, 1–2).

For the most part such terminological distinctions, along with the others we shall examine, are settled today. "Coder" is now equivalent to "programmer," while the newer terms "software engineer" and "software architect" have emerged to denote the planning and design responsibilities preceding and exceeding implementation that both Adams and colleagues and Hopper had associated with "programmer" and "programming" alike. Terminology for describing the programming process and tools is even more firmly disambiguated. The software *compiler* of a compiled programming language transforms human-written and human-readable source code in multiple phases, invoking a *preprocessor*, a *compiler*, an *assembler*, and a *linker* in turn. *Assembly* refers to the transformation of code written in a so-called higher-level programming language into a hardware-dependent *assembly language* specifying mnemonic operation codes and other instructions, and subsequently (or directly) into a symbolic *machine code* representation of binary logic. *Compilation* refers to the entire process of preprocessing source code, assembling it in this sense, and linking a binary executable to necessary libraries in an operating system context. *Execution,* a distinct phase in the life cycle of a program, follows compilation and involves running the program on hardware. In this phase, a CPU interprets instructions, accesses memory, and may interact with input and produce output until a run is complete or encounters an error.[8]

Both interpreted and, increasingly, compiled programming languages can be *interpreted* by a software *interpreter* that provides the programmer with a REPL (Read-Evaluate-Print-Loop) interface for writing and evaluating statements without manually compiling a source code file. The terms *scripting* and *byte compilation* are associated with interpreted languages and some compiled languages that are interpreted by so-called virtual machines, while source-to-source compilers, also called *transcompilers* or *transpilers*, transform source code in one higher-level programming language into source code in a different higher-level programming language. It is only these last few, much more recently developed terms that display any of the ambiguity or polysemy of the words "programmer" and "coder" as they were used in the 1950s; as I shall suggest, it is here that the translation metaphor that David Nofre, Mark Priestley, and Gerard Alberts identify as central in early computing retains the links to linguistic concepts that were obfuscated or discarded in the constitution of a specialized lexicon for source code compilation.[9]

Translating Translation

For this reason among others, we might come to see a place for the history of computer programming languages in the much longer and broader history of literary and scientific imaginations of a universal language. As chronicled by Umberto Eco in *The Search for the Perfect Language in European Culture* (1995), the latter history can be understood to comprise three phases. The first is the imagination, then the construction of artificial primary, auxiliary, and cryptographic languages, from the late seventeenth through the late nineteenth centuries in Europe and Asia. Second, worldwide national language standardization, alphabetization, and writing reform, from the mid-nineteenth century into the twentieth. And third, the global and total simulation of writing (and later, speech) in digital code, from 1945 to the present.

While Eco's book is typical in its entirely glancing and truncated treatment of the computer programming languages of the twentieth century, one might lament equally the self-imposed constraints on existing histories of programming languages by Jean E. Sammet (1969), Donald E. Knuth and Luis Trabb Pardo (1976), Richard L. Wexelblat (1981), and Thomas J. Bergin and Richard G. Gibson (1996). Irrespective of—indeed, in some ways perhaps precisely *owing* to—their technical sophistication, such work might be said to have documented the linguistic history of computing without reflecting very extensively on its broader global historical, cultural, and political contexts as a history of human uses of language.[10]

Knuth and Pardo wrote of the "pre-Babel" days preceding the release of the first version of the Fortran language and the "explosive growth in language development" that followed from 1957 on (Knuth and Pardo 1976, 2, 95). The gentle irony of this remark reminds us that the history of programming languages up to 1957 was anything but peaceably unilingual (or monological) in character: as early as 1954, Saul Gorn noted

that despite aggressive university investment in computer design, construction, and use for data processing, universities were "reluctant to
train programmers, feeling that there are too many specialized codes."
Perhaps, he went on to suggest, "this reluctance will vanish if we can
provide a code more or less independent of the machine" (Saul Gorn
1954, 75). The specificity and thus the incompatibility including lack of
interoperability of machine code instruction sets provided by hardware
processor manufacturers was perceived as a problem almost from the
start, and a rich discourse developed borrowing concepts of language
and translation from a linguistic lexicon.

Brown and Carr (1954) remarked that many of the earliest programmers had clung to direct binary or octal coding of instructions, resisting
even the use of decimal notation (Brown and Carr 1954, 85). But as Saul
Gorn's introductory remarks to UNESCO's International Conference
on Information Processing in Paris in 1959 make clear, by the time that
Brown and Carr had written "Automatic Programming and Its Development on the MIDAC," a range and variety of general-purpose computers
were being used the United States. "At Aberdeen Proving Ground [the US
Army facility in Maryland]," Gorn reflected,

> the Eniac, Edvac, and Ordvac [computers] were in use simultane-
> ously. This prompted a "universal coding" experiment in which a
> simple, general purpose, common pseudo-code was used as an input
> to both Edvac and Ordvac. The experiment separated the processes
> of "assembly" and "translation." It showed that the assembly process
> was independent of the machine, and could be achieved within the
> common language by an automatic translation which had the prop-
> erty of leaving the assembled code in the common language. (S. Gorn
> 1960, 117)

Describing the development of early "Input Translation Programs"—
"programs which translated both numbers and instructions from an external decimal language more useful to human beings, into an internal
binary language more useful to the machine"—Brown and Carr advocated
"recognition of the principle of a dual language system and a programmed
translation between the two languages, using the computer to perform the
translation" (Brown and Carr 1954, 85). Whether performed upon code
written in decimal notation, in mixed alphanumeric and decimal notation, in algebraic alphanumeric and symbolic notation, or in abbreviated

English words (or on combinations of all of these), the word "translation" was used to describe two different processes, each of which might be performed separately, but which might also be linked as phases on an installation that could handle both of them. The first was what Brown and Carr called "pre-translation," by which they meant the compiling of a program as preprocessing (rather than as a complete or final step). Compilation-as-translation might, for example, include the assignment of absolute physical memory addresses to the abstract symbolic or floating memory addresses specified in the source code. The second was "running-translation" (a phrase they wrote in hyphenated form), or the "interpretation" of a program, which performed its instructions without as much symbolic preprocessing. Compiling, they wrote, was possible on systems with significant storage capacity (for example, which were equipped with magnetic tape storage), while interpretation "was the only feasible translation method" on installations equipped with smaller amounts of storage (Brown and Carr 1954, 86). It was compilation-as-translation that made possible the next and final step, the "automatic assembly" of a final program.

Brown and Carr noted, however, that while they made programming less fundamentally arduous, such "external languages" including decimal notation and mnemonic instruction codes could not entirely prevent programming mistakes, whether "arithmetic, logical, or clerical" in nature—and that they might actually produce new ones. In fact, "the opponents of translation procedures," they reminded their readers, had objected that the use of a dual-language system with a translator only increased the burden on the programmer, since in order to check the program for mistakes the programmer would have to "know and use both external and internal languages, which nullifies some of the advantage" (Brown and Carr 1954, 86–87).

Such "opponents of translation procedures" were very quickly shunted aside. "By 1955," Gorn concluded in his introduction to the UNESCO International Conference on Information Processing, "it was clear to all who were deeply involved in automatic programming, or other non-numerical data-processing, that we were dealing, not with computers, but with general information transformers, or symbol manipulators. The approach to communication with these machines had become frankly linguistic and syntactical" (S. Gorn 1960, 117). Gorn was consistently both aggressive and grandiose in his pronouncements, which are marked by the specifically technocratic cosmopolitanism and applied-scientific triumphalism of the immediate postwar period. "Our civilization," he wrote in an earlier essay titled "Planning Universal Semi-Automatic Coding,"

> is characterized by a herculean effort to control our physical environment. In order to do so our arts, sciences, and professions have developed a profusion of specialties either to deal with the sharpening of our methods of communication or to apportion the tons of information in such a way that single individuals can have them at their command. If this civilization is not to become a Tower of Babel of specialist languages, its vast tonnage of information must be processed mechanically, whether the processing is computing or translating. The artists, scientists, and professionals can then return to a common language and creative thinking. (Saul Gorn 1954, 74)[11]

"Analogies work both ways": Language and/as/or Code

I have already noted that the first and most important requirement of such a common language, or "universal code," was not expressivity, that quality programmers most prize today, but translatability. Translatability was not necessarily imagined as an intrinsic quality of a particular language design; rather, it might mean merely that routines, either in formal description or actual implementation, had been developed for translating hardware-independent universal code into hardware-specific operation codes or machine instructions (Saul Gorn 1954, 75). In fact, in their notes from the 1954 MIT summer session program on the topic "Digital Computers: Business Applications," Adams, Gill, and their colleagues explicitly suggested the linguistic-communicative concept of translation as an alternative to the explicitly behavioral concept of virtualization that dominates our own age of "cloud" or distributed computing as a service today. "It has been stated," they wrote,

> that the net result of automatic coding techniques is to cause one machine to imitate another. Alternatively, they may be thought of as allowing the programmer to write his programs in a different *language*, one that is translated to the machine's internal language by the processing routines. (Adams and Gill 1954, 16-6)

The translation metaphor also appeared in reflections on the relationship of a universal code to other notation systems in which mathematicians and engineers formulated problems they hoped to program computers to help them solve. In the imagination of a universal code as a third language or interlingua, for example, the translation metaphor could grow baroque and strained:

> The crux of what has been done in the past has been the introduction of a third language into programming—the first two being the language of the machine and the language in which the problem is formulated. [. . .] Until very recently programmers have been like an American who can speak German who finds himself with a Frenchman who can speak Russian. In order to communicate with the Frenchman, the American must find a German who can speak Russian. It would be simpler if the American would learn to speak French or the Frenchman English but, of course, the American would rather have the Frenchman learn to speak English than learn French himself. Similarly, programmers would like computers and data-processing machines to understand the language in which their problems are formulated. (McGee 1957, 57)

Other formulations drew not on the cosmopolitan European union of the postwar order, much less on the nascent era of decolonization, but on earlier paradigms for the civilizing mission of empire:

> Sending a problem into a computer nowadays is like sending an expedition to Africa to trade with the natives. It has to be complete with the missionaries to translate and possibly convert the natives. If the missionaries speak only French and the native tongue then we must speak French but if the missionaries speak English too then everything is all right. (Wegstein 1956, 6)

The use to which such borrowed concepts were put in early computing is at odds with contemporary linguistic translation theory in so far as the latter, along with much contemporary cultural theory more broadly conceived, has since the 1960s come to reject any analogy binding code and language, such as that provided most succinctly by Earl J. Isaac in 1952: "Coding for digital computers is a process of translating from one language to another" (Isaac 1952, 117). The transformations involved in the compilation of source code require, indeed presume in the most uncontroversial sense, what the translation studies scholar Anthony Pym calls the "paradigm of equivalence" between the values denoted by one symbolic system and those of another into which it is to be transformed (Pym 2014, 7). There would be no point in and no motive for developing assembly codes or a higher-level programming language if no mechanism could be created to literally map, more or less directly and by equivalent values, both such symbolic systems to the machine code representing the execution of instructions.

This is not to say that natural linguistic translation is never imagined in terms of such relations of equivalence; on the contrary, as Pym notes, it is "close to what many translators, clients, and translation users believe about translation" as a purely or principally practical activity, delinked from philosophical and political questions of language, history, geography, and human community (Pym 2014, 6). But it does remind us of something demonstrated from the start by early research in machine translation, whose development runs very closely alongside the earliest history of programming. That is that the domains in which such relations of equivalence obtain and can be practically mechanized or computerized are carved from the domain of natural language by the standardization of jargon: that is, the specialized and above all deliberately precise terminology used for communication in the sciences, engineering, and almost any contemporary profession, not excluding the professionalized practice of research and teaching in the humanities. Insofar as it is disproportionately idiomatic, the everyday colloquial use of any natural language is too imprecise for such service. And even though it communicates its results in a standardized jargon like any other, when linguistic translation theory has rejected analogies binding code to language, it is in the name of such idiomatic usage on the one hand and of literary language on the other. It is these latter domains which linguistic translation theory took as its research objects, beginning in the 1960s, in part precisely *because* of the remoteness both domains maintain from the precision of a standardized jargon.

Where the concerns of linguistic translation theory have been most remote from the foundational assumptions of the computer science discipline of programming language theory (PLT), these two domains are for all practical purposes incommensurable, and there is nothing particularly wrong with that. That incommensurability does suggest that the attempt to bridge it, by resort to the translation metaphor in early computing, was exactly but also *only* that: a metaphor, that is, a transfer or carrying over of one thing to another across an at least potentially resistant distance. In an era when economic suffering and solutionism alike encourage us anew to conflate knowledge of computer programming with basic literacy in a natural language, it is worth reminding ourselves that the translation metaphor in early computing had a short life for a good reason.

Jörg Pflüger has argued that dealing sensibly with the language analogy in computer science *requires* us to distinguish human natural languages from the formal languages of computer science, and that while computer science does indeed have a great deal to do with "language" broadly speaking (that is, its domain is not exclusively mathematical), its contact with the domain of human language is achieved exclusively through metaphorization.[12] And if one is inclined to argue that natural language itself is already metaphorization, then what we are talking about is a second-order or supervening metaphorization or "metametaphorization." This dynamic is no mere curiosity: it can be embedded in an economic-political context, such as that marked by the drive for "code literacy" I mentioned above, in a manner that makes it not only interesting, but urgent to study. As Pflüger puts it:

> As computer technology deals not only theoretically with formal concepts but also reconstructs reality with its models, there exists a reciprocal effect between mechanization and language. Analogies thus work both ways. The artifacts of computer science affect, in turn, the realm of language, where the mind is at home, and they organize social reality in analogy with reductionistic models of language. One effect of computerization is that many activities are (re)structured in such a way that they seem to be organized by a "grammar of action" within a framework of formal languages. (Pflüger 2002, 130)

In that context, it is fair to say that the analogy chosen by Joseph H. Wegstein in 1956, that "sending a problem into a computer nowadays is like sending an expedition to Africa to trade with the natives," unintentionally but also unreflectively extends the language metaphors of a nascent computer science into the association with colonial violence from which the national philology of the age of European empire, the main progenitor of the linguistic and literary humanities today, has still not completely extricated itself. Would it be equally fair to observe that the close association of the translation metaphor with automation extends it into an association with endocolonization, that postwar dynamic in which the products of publicly funded military innovation are dispersed in both the civic sphere and the private sector in patterns slowly but steadily incrementing both the disenfranchisement of labor and the militarized policing that tries to keep a lid on it?

Brown and Carr had opened "Automatic Programming and Its Development on the MIDAC" with the following words: "A new term has developed recently in the automatic digital computer field. 'Automatic programming,' a concept which 10 years ago would have been worthy only of the science-fiction writers, is now within view, if not reach, of the users of high-speed digital computers" (Brown and Carr 1954, 84). If the euphoria this sentence clearly suggests was modulated in Stanley Gill's remarks on the same topic, it was with the pragmatism of an engineer who aims not for a complete solution but only for a minimization of the problem. "We must bear in mind," Gill wrote, "that there will always be two phases in the preparation of a program for a computer: the human phase and an automatic phase. Automatic programming techniques are designed to reduce the human phase at the expense of the automatic phase" (Gill 1954, 98).

For her part, Grace Hopper had closed "Automatic Programming— Definitions" (1954) by calling for the development of a "universal" pseudocode for which every computer installation would provide its own "interpreter or compiler" (Grace Murray Hopper 1954, 4). Remarking on the accumulation of code for existing computers and the proliferation of new computer designs using new sets of hardware instruction codes, she also called for "translators to make available 'old' coding to new computers," to conserve programming time and labor. "Since this will be a mechanical operation," she concluded, "we shall look at the computers to do this job for us." More generally, she suggested, "just as compilers control generators to produce programs to process data, we shall soon be talking of systems containing computers controlling and directing computers" (Grace Murray Hopper 1954, 5).

It makes sense to speak of automation as recursive whenever a system is designed to automate a process that involves automation itself. In other words, the automated process is designed to repeat or replicate itself, creating a self-referential or self-replicating loop. For example, imagine a system that is designed to automatically generate program code based on a set of input parameters. The system may be designed to use machine

learning algorithms to analyze the input data and generate code that performs a specific function. However, if the generated code itself includes
automated processes, then in context those processes may be thought
of as recursive in structure, in a generalized sense. Another basic example of recursive automation, perhaps even easier to grasp, is the use
of robotic systems to automate a manufacturing process. Such a system
might be designed to perform a specific set of tasks, such as welding or
painting, but it may also be programmed to adjust its own settings or
perform maintenance tasks on itself. This is a recursive activity in which
the system is both performing tasks and optimizing its own performance
(for better or worse).

In mathematics, recursion refers to the definition of a sequence or
function in terms of itself, ensuring that the value of a given term in the
sequence or function is defined in terms, as it were, of previous terms.
Similarly, in computer science, recursion refers to the technique of defining a function in terms of itself—but it also means calling a function,
or executing its instructions, from within itself. This allows a function
to repeat its assigned instructions multiple times in sequence. In theory,
in both mathematics and computer science, recursion is a self-referential
procedure that could be repeated indefinitely: for example, in a numerical or geometrical sequence like the so-called Fibonacci sequence or the
Cantor set, which have no meaningful limits. In practice, in the development of software, a recursive process must be provided with limits. Each
time a function calls itself, it uses additional memory to store the current
state of the function, including its arguments and local variables. Given
enough function calls, memory usage will expand to the capacity available, causing a stack overflow or other memory error. Processor usage
may also expand to the capacity available, with material consequences
in the form of heat damage.

At the same time, the examples given above, of a code generation
system and robotic manufacturing, are merely extrusions of the fundamental technical logic of software itself, regardless of its specific applications. In a fundamental sense, a software application is not the *product*
of a programmer's programming as much as it is the *automation of the
programming* that would otherwise have to be done in real time, to perform a computational task. The very concept of a stored program, meaning a program stored in memory so it can be used later, expresses this
fundamental technical logic of software. When programs are stored in

memory, and thus treated as data that can be manipulated by other programs, a computer can execute different programs without needing to be physically rewired or reconfigured. (Famously, the ENIAC computer at the University of Pennsylvania, whose core components were vacuum tubes, needed specific wiring for each calculation, while the somewhat less famous Mark I at Harvard University was initially reconfigured by repositioning mechanical dials and switches.) There can be no general-purpose computing without this fundamental and intrinsic automation.

Conclusion

Janet Abbate has argued that an early ideal of automatic programming was pursued by both "programmers who [. . .] hoped to liberate themselves from drudgery and empower themselves to do more interesting work" and managers who "hoped that the new methods would deskill, discipline, or eliminate programmers," and that gender "played an important though subtle role in these developments." Abbate rejects arguments made by Joan Greenbaum and Philip Kraft, Michael Mahoney, Paul Ceruzzi, Nathan Ensmenger, and others who have suggested that the priorities of managers dominated, in these developments. Arguing that automatic programming "does not fit the classic deskilling model," Abbate suggested that in relieving the programmer of the drudge work of programming, the automation provided by software compilers also contributed to the masculinization of programming as a profession, with automatic programming tools displacing some of the rote labor to which many if not all women in a nascent industry had been confined, to that point, while elevating the new software "engineering" role men invented for themselves.[13]

Abbate sees an important difference between the arguments that Grace Hopper made in 1954 and those that appeared two years earlier in "The Education of a Computer," delivered at the May 1952 meeting of the Association for Computing Machinery. Hopper's "brash" earlier confidence that "a computer could replace the human brain" was absent, Abbate argues, from the 1954 paper, in which Hopper "decisively rejected the notion that the programmer would be replaced by a machine" (Abbate 2012, 82–83). But it seems both possible and fair to interpret Hopper's continued speculations in 1954 as something less than a decisive rejection of that notion; and just as we can say without controversy that Hopper herself contributed a great deal to the masculinization of programming, as Abbate describes that process, we cannot conclude that Hopper "never fully adopted a management perspective" *because* (the word Abbate chose is "since") "women had no realistic hope of rising into the upper ranks of computer firms" (Abbate 2012, 82). Programming was (and remains) the privileged, explicitly agential form of a broader automation of data processing, itself very much on the agenda, as we have seen (and as Abbate also shows), by 1954; in this context, programmers' control of that privilege does not mean that they never work against their own interests. The evidently strong concern of male programmers to distinguish a *new* masculine role for themselves, in such a context, speaks to the breadth of both the determinations and the contradictions at hand. Along with gender, other determinants of labor status, such as race, raise additional questions about the pursuit of recursive automation, considering that the labor implications of *Brown v. Board of Education*, decided by the US Supreme Court in May 1954, cannot have been far from any thinking person's mind at the time. Finally, there is also the question of whether something in the fundamental technical logic of programming itself, rooted as it is in an ideal of universal translatability, produces further automation irrespective of the consequences.

I suggested earlier that when the "bottleneck" of the human operator, as Adams and Gill imagined it in 1954, is backlit by the history of computer programming, it seems to shadow some problems of labor and economic organization we have struggled with much more recently. Recent years have seen the emergence and growth of initiatives to add instruction in software programming to primary and secondary educational curriculums in the United States and the United Kingdom, as well as to provide such instruction at no cost or low cost via internet, outside the historical institutional boundaries of the educational sector. This is a complex and multifaceted development, whose dimensions it was not my purpose to explore fully here. But it is unquestionably one of the political and cultural effects of the developments that began with the 2007–8 US housing and banking crisis, which briefly offered Silicon Valley the opportunity to rebuild the reputation tarnished by the 1997–2000 dotcom bubble and collapse, sent institutional investors fleeing the housing market in search of profit in higher education reform, and encouraged policymakers, managers, and workers to seek quick fixes to unemployment through the technical retraining of an unemployed and underemployed managerial middle class.

A 2013 report by *Harvard Business Review* titled "America's Incredible Shrinking Information Sector" has doused such expectations, noting that far from having grown, "the information industry [. . .] shed more jobs in the first decade of the millennium than any other sector except manufacturing," and that "the culprit, ironically enough, is tech-driven innovation, which has produced dramatic gains in efficiency and widespread automation" (Robison and Chang 2013). One might have wondered if similar observations had prompted recent changes in the US Bureau of Labor Statistics *Occupational Outlook Handbook* for the category "Computer Programmer." Whereas the 2014–15 edition of the *OOH*, published in January 2014, predicted annual job growth of 8 percent for the period 2012–22 (which it noted was "as fast as average" rather than higher or lower than average), the 2016–17 edition published in December 2015 predicted an annual decline of 8 percent for 2014–24, amounting to a loss of 26,500 jobs, while the December 2020 update predicted an annual decline of 9 percent for 2019–29, amounting to a loss of 20,100 jobs. From December 2015 until September 2022, the explanation provided by the *Handbook* for this prediction, which has been repeated each year since 2016 and is still current as of this writing, was the outsourcing of programming jobs: "Computer programming can be done from anywhere in the world, so companies sometimes hire programmers in countries where wages are lower. As a result, employment growth for computer programmers in the United States may be limited." But the September 2022 update provided a substantively different explanation:

Computer programming work continues to be automated, helping computer programmers to become more efficient in some of their tasks. Many companies are leveraging technologies to automate repetitive tasks, such as code formatting, to save time and money. Automation of this routine work could allow computer programmers to focus on other tasks, such as strategic planning activities, that cannot be automated. In addition, some computer programming tasks are more commonly done by other computer occupations, such as developers or analysts.[14]

Of the earliest history of "automatic programming" that I have reviewed here, Nathan Esmenger has observed that, far from eliminating the software industry's need for programmers who were highly skilled, well trained and well compensated, and thus difficult to manage, higher-level languages "contributed to the elevation of the profession, rather than the reverse, as was originally intended by some and feared by others" (Ensmenger 2010, 84). Though that paradox and apparent confirmation of Say's Law has persisted, at least through Silicon Valley's major boom cycles thus far, nothing guarantees its perpetuity, and the trajectory projected by the *Harvard Business Review*'s report cannot be obscured. "For years," Bix (2002) has observed, "managers [had] appeared relatively immune" to technological unemployment through automation, right up to the 1980–81 recession, which "saw three layoffs of blue-collar workers for every one white-collar layoff. The years 1990 and 1991, by contrast, brought one downsizing in management, professional, or clerical staff for every two blue-collar reductions" (Bix 2002, 281). Recognizing the specific role played by the accumulating layers of software automation, during the last decades of the twentieth century, Bix argued straight-forwardly that "in the 1990s, development of increasingly sophisticated computer applications ripped a hole in the comforting assumption that a specialized education would protect a person from technological unemployment" (Bix 2002, 289).

In other words, this dynamic is now forty years old. Bix also noted that "awareness of the technological unemployment issue rose and fell with the nation's overall economic fortunes. By mid-1997, as joblessness fell to a twenty-year low, the fear of displacement voiced three years earlier by [Stanley] Aranowitz and [William] DiFazio, two years earlier by Jeremy Rifkin, just the previous year by *The New York Times*, suddenly receded. A soaring stock market, low unemployment, and low inflation once again pushed back concern about the social implications of workplace change" (Bix 2002, 302).[15] It is true that as Bix also observed, the intra-elite "class warfare" predicted by Jeremy Rifkin in *The End of Work: The Decline of the Global Labor Force and the Dawn of the Post-Market Era* (1995) did not materialize: "the 1990s had not brought a massive wave of guerrilla attacks on Silicon Valley CEOs and engineers by those who felt excluded from its best jobs" (Bix 2002, 300). Nonetheless, "reports on the modern militia movement suggested its leaders had expanded membership by tapping into the vulnerability and anger of economically strapped Americans, especially those suffering from the decline of blue-collar jobs" (Bix 2002, 300). This observation remains entirely, even extravagantly pertinent in the era of US political history that began in 2016.

This is not even to mention the historically unprecedented technology industry layoffs of 2022–23. At the time of writing, the most spectacular of many quite spectacular numbers remains the 18,000 corporate positions that Amazon announced it would eliminate in early 2023: what Jason Del Rey, writing for *Vox,* called "by far the largest total number of job cuts in its history." "To put the abruptness of these changes in context," Del Rey wrote, "as recently as June 2022, Amazon's career site had listed more than 30,000 job openings—that's not a misprint—in software development positions alone. But by mid-January, it had fewer than 300" (Del Rey 2023). With such past, present, and future developments kept properly in mind, might we conclude that in the end, the so-called code literacy promoted by policy wonks, ed tech consultants, and higher ed reformists and freelance opinionistas of all stripes, for most of the 2010s, is nothing like basic literacy at all—that is, a modern basic need, widely regarded as a fundamental human right—but rather a self-eliminating condition with little future as a human labor skill in the United States or any other economy? Might it even suggest that learning to code is a red herring, an imagined substitute for the more difficult and permanent human effort of economic regulation that in the aftermath of the 2008 collapse might have produced something more than austerity and its protraction of suffering (and for that matter might have prevented the collapse of 2008 to begin with, had it not been left undefended)? Have the "fun work" and the imagined meritocracies of a cyberlibertarian Silicon Valley startup and coding culture gradually displaced commitments to the distribution of resources and opportunity across intersections of class, race, gender and sexuality, ability, and geopolitical space? Such outcomes ought not to surprise us, if we understand that the history of computing in general and of computer programming in particular is a history of recursive automation, that is, of the addition of successive layers of control through which higher- and lower-level codes are "translated" to each other, and that has as its vanishing point, at once by design and by accident, a historical moment when human control of this new kind of writing recedes.

———

Thus far, we have considered the translation metaphors employed by the first programmers of electronic computers as they both borrowed and invented terms for describing the writing and execution of programs. As we have seen, before such technical categorical terms as "higher-level" had settled into common usage, early programmers borrowed from the national and comparative philological domain of natural or human languages to describe relationships among activities and processes in computing. Words like "foreign," "native," "translation," and "translatability" were widely used at this time alongside or in combination with new terms from an emerging technical lexicon. I have suggested that the most useful way to understand the procedures denoted by the term "translation," in such a context, is to consider them *automations of automation:* that is, self-extensions, or extensions into or onto itself, of the fundamental purpose and the basic technical logic of software as automation. To provide this insight with a philologically broader context, chapters 2 and 3 of this book are focused on contact zones within a computer program where natural languages and programming languages appear to touch or intermingle.

TWO

Really Reading the Code,
Really Reading the Comments

At the beginning of chapter 1, I discussed the 1953 summer session at the Massachusetts Institution of Technology, for which Charles W. Adams taught the first of a sequence of programming courses. I also mentioned "Digital Computers: Business Applications," one of the first substantive documents to emerge from Adams's summer teaching. From the beginning, I suggested in analyzing the discourse of Adams and his colleagues in that report, it was recognized that the technical logic of computer programming was a logic of automation, indeed a logic of self-automation or even recursive automation. In the present chapter, I suggest that a syntactic feature common to all programming languages used today serves as a register of this logic of automation, leaving traces in code that are legible if one knows where to look. That feature is the comment.[1]

Comment/Code

By "comment" I mean natural-language text within a computer program that, unlike the code accompanying it, is not written for computational interpretation or compiling into operation codes and/or machine instructions: that is, which is not written for program execution, but rather exclusively for reading by a (human) programmer. The so-called reserved words or keywords of every single one of dozens of major programming languages used in software development from the 1950s to the present have been, and will almost certainly continue to be, words or abbreviated forms of words in the English language—but they are English words whose constituent symbols are parsed by a compiler and read with comprehension by a human programmer. The inclusion in some programming language designs of noncompiled "noise words" that in modifying and/or joining reserved words provide a syntactic flexibility that is imagined to extend toward natural language (examples include Cobol, HyperTalk, Applescript, and Lingo) may have expanded that English lexicon in specific cases, but they have never challenged it. There are, however, only one or two reserved words or symbols used to demarcate program comments, in any programming language, and anything following that character belongs to another domain, insofar as a compiler ignores it. In articulating the categorial distinction between nonexecutable and executable program text as a distinction between natural language and code, Letha H. Etzkorn and colleagues' unobfuscated definition of program comments as "natural language, textual descriptions of the software" is especially useful, at once terse and true (Etzkorn, Davis, and Bowen 2001, 1732).

The second part of their definition has equal importance. "Comments," they write, "are located within the computer software, typically very close to the computer code that they describe" (Etzkorn, Davis, and Bowen 2001, 1732). The apposition and adjacency of comments and code, as articulated by the entirety of Etzkorn and colleagues' brief but pithy definition, should be understood as sites of difference or differentiation between, rather than conflation of, natural language and code. Program code and program comments are more different than they are similar, though they are found in the same textual artifact imbricated or interleaved with each other.[2]

One might be tempted to describe program comments by analogy to the footnote or endnote in paginated printed text, or to its manuscript antecedents, from the simple gloss to the scholium and commentary, or to a descendant, the hypertext link imagined as text encoded with (among other elements) instructions for an executable jump from one address to another in stored text data.[3] Or, differently, to such epiphenomena of book reading as titles, title pages, dedications, and epigraphs, which Gérard Genette catalogued as components of "le péritexte éditoriale." Similarly, one might consider the relationship of program comments and code by analogy to that between a book publisher's public "epitext," meaning the marketing and promotional texts supporting the published book as a product, to the private epitext of such documents as diaries, letters, and anticipatory or early versions that precede (or follow) publication.[4] Tempting as they are, none of these approaches is appropriate for describing program text, which is very rarely handwritten and seldom read from a printed page, at least relative to the frequency with which it is read from a display screen; which does not conceal the articulation of a "jump" beneath an interface, but articulates it directly; and whose dependent components are integral, rather than either peritextual or epitextual in Genette's senses of those terms. The concept of transcluded text, as imagined (and thus far unsuccessfully implemented) by Ted Nelson, may come closer than any of these other forms to capturing the apposition and adjacency of comments and code.[5] But even here, it can be observed that by design, transcluded text identifiably combines identifiably different sources, whereas comments have no other source than the source code document in which they are found. Etzkorn and colleagues' technically correct *and* responsibly historicized definition has the great merit of restraining such negotiations.[6]

Etzkorn and colleagues go on to argue for the understanding of program comments as a professional technical "sublanguage of English" like the jargon of research disciplines and trades and professions. Though their analysis of the specific characteristics of this restricted sublanguage—its contraction to a small subset of English verbs, its elision of grammatical articles, personal pronouns, and moods other than the present indicative or imperative—leads them to speculate about the automated (that is, "static") analysis of program comments using computational techniques for natural-language processing, they note that these same grammatical characteristics make comment text a nontrivial challenge for automated analysis.

Technical research on program comments tends to focus more on the concept of documentation and the document than on the sociolinguistic issues considered by Etzkorn and colleagues. In an early example of such research titled "Some Comments on Comments," Jon Sachs distinguished among three forms of documentation used in software engineering: the external documentation written for users of a software application, the operational documentation written for technical service providers serving users, and the internal documentation used by programmers themselves. External and operational documentation may or may not be produced by programmers (they may instead be produced by a professional technical writer whose primary task is to write documentation of an application, not to create or maintain it), but internal documentation is produced exclusively by programmers, for programmers.[7] And whereas the readers of external and operational documentation read it in a context deliberately separated from any reading of the program code itself (in a manual, for example), the same is not true of program comments, which are composed alongside program code. Comments, Sachs suggests, are unique not only in their intended readership and mode of address, but also in their situation: "they are actually a part of the object they document" (Sachs 1976, 7).

Others have imagined source code not as an internally documented object, in this sense (a sense that preserves the distinction between comments and code), but as "a document, describing the program you are creating" (Goodliffe 2007, 25). In this formulation, which elides the distinction between comments and code, the document in question is program text taken as a whole—a whole that might in fact not need to include any comments, if the program code were sufficiently readable on its own: "The only document that describes your code completely and correctly is the code itself" (Goodliffe 2007, 59). I will return to this bifurcation of approaches below, since it structures what technical discussion is available in the documentation of software engineering practice, specifically project (but also personnel) management, and represents two meaningfully contrasting concepts of the text of a program: one best understood as an idealization of execution that rests on an analogy between the human reading of program code and its error-free compilation, and another, broadly "documentary" approach that accepts or even requires that a program maintain two intractable, even incommensurable textual dimensions, both of them human-readable but one necessarily also nonexecutable.

The Elements of Programming Style

Today we use the phrase "higher-level programming language" to refer to the so-called third-generation notation systems that emerged in the mid-1950s combining algebraic expressions with English-language keywords. The principal motive for designing so-called higher-level languages was to reduce the difficulty of direct hardware programming in a first generation of purely numerically represented instructions, or "machine code," and a second generation of equally hardware-specific operation codes represented by mnemonic abbreviations (today called "assembly language"), though their creators quickly understood the great potential benefit of being able to run the same program on more than one computer of different hardware design, as well. The first implementation of what we now call a higher-level programming language was Short Code, designed (and originally named Brief Code) by John W. Mauchly in 1949 for the Eckert-Mauchly Computer Corporation's BINAC and implemented by William F. Schmitt for the BINAC and subsequently for the UNIVAC I. In a brief memoir that is the only surviving document of Brief Code/Short Code I, Schmitt makes no mention of comment syntax in the BINAC and UNIVAC I versions that he implemented in 1949 and 1950, or in Short Code II, implemented for UNIVAC II by Albert B. Tonik and J. Robert Logan in 1952.[8] Neither the Speedcoding system created by John Backus for the IBM 701 in 1953 nor the system devised by J. Halcombe Laning Jr. and Neal Zierler for the Whirlwind computer at M.I.T. (1952–53) appears to have included any facility for program comments.[9]

The first definition of a syntax for program comments in a higher-level language can be found in *The Fortran Automatic Coding System for the IBM 704: Programmer's Reference Manual*, written by David Sayre and published in 1956 by IBM's Applied Science Division and Programming Research Department as a reference for what would come to be called Fortran I, the first specified version of the language.[10] "Punching the character C in column 1" of a Fortran coding sheet, the *Programmer's Reference Manual* explained, "will cause the card to be ignored by FORTRAN. Such cards may therefore be used to carry comments which will appear when the [card] deck is listed" (*The Fortran Automatic Coding System for the IBM 704: Programmer's Reference Manual* 1956, 8). The specifications for what we now call Algol 58 (1958), Cobol 60 (1959–60), and Lisp 1.5 (1962), three other languages developed in the 1950s that, along with Fortran, either achieved widespread adoption or significantly influenced subsequent language design, each included a comment syntax.[11] In each case, program comments were given a definition that has remained stable and is consistent with usage today, though the Cobol 60 and Lisp 1.5 specifications are as laconic as that of Fortran I, with the authors of the Algol 58 report being the first to articulate a categorial distinction between program comments and code. "Comment declarations," they wrote, "are used to add to a program informal comments, possibly in a natural language, which have no meaning whatsoever in the algorithmic language, and no effect on the program, and are intended only as additional information for the reader" (Perlis and Samelson 1959, 55).

Along with the cryptic **REM** (for "remark") of BASIC, the early Algol and Cobol reserved words **comment** and **NOTE** are long gone, and program comment syntax today is defined almost exclusively by punctuation: the forward slash or combination asterisk and forward slash in the so-called C family of languages;[12] the hash symbol (octothorpe) in scripting languages like Perl, Python, and Ruby; and the semicolon in Lisp dialects and languages based on Lisp. Some languages permit the programmer to insert both line comments, limited to a single line of text beginning either at the margin or following the end of a line of code, and block comments, which may span multiple lines. Languages like Java, Python, and Lisp-derived languages, among others, allow for a third type of comment called documentation comments or documentation strings, typically inserted on the first line of a unit of code such as a package, class, method, or function and processed by automated documentation generators. Even in the latter case, however, the demarcators of such comments are typographic, rather than lexical. Arguably, the reduction of comment syntax to typographic punctuation symbols marks increasing definition in the symbolic enclosure of all nonparsable natural language in program text: purely typographic indicators such as the hash

C ← FOR COMMENT / STATEMENT NUMBER	CONTINUATION	FORTRAN STATEMENT	IDENTI-FICATION
C		PROGRAM FOR FINDING THE LARGEST VALUE	
C	X	ATTAINED BY A SET OF NUMBERS	
		BIGA = A(1)	
		DO 20 I = 2,N	
		IF (BIGA - A(I)) 10, 20, 20	
10		BIGA = A(I)	
20		CONTINUE	

FIGURE 3. Image of Fortran coding sheet, as depicted in *The Fortran Automatic Coding System for the IBM 704: Programmer's Reference Manual.* Reprint courtesy of IBM Corporation © 1956.

symbol and repeated forward slash or semicolon act as visual barriers between two domains of text. While it is still common for a programmer to write a first draft of a program in pseudocode, or nonexecutable text expressing the logic of a solution while leaving its implementation for later, such a pseudocode draft is almost always deliberately composed within a comment block or as a series of comments, so as to be ready at hand for, even interleaved with, the program's implementation.

The canonical treatment of programming style is Brian W. Kernighan and P. J. Plauger's *The Elements of Programming Style* (hereafter *Elements*), a slim volume originally published in 1974, a moment when the so-called structured programming revolution had reached its maturity. The most common form of higher-level program text in the late 1950s and early 1960s was a plain sequence of instructions incorporating so-called "go to" statements that switched execution from one line of the program to another in the manner of jump statements in assembly-level operation codes. Emerging from struggles with the difficulty of reading, understanding, and modifying such programs after they were written, structured programming emphasized use of the compound statements or block structures introduced in Algol 58 and subsequently incorporated into other languages.[13] As a product of this shift in practice, *Elements* articulated standards for program code readability as a measure of the arrangement and organization of code itself, rather than as a measure of supplementary (internal or external) documentation, and it emphasized the legibility that might be achieved by modularizing, encapsulating, and otherwise arranging program code in compound forms before one even began to document it in comments.

In their emphasis on the discipline of a structured programming style over the production of documentation, Kernighan and Plauger articulated a complete set of principles for self-documenting program code—that is, program code that can be fully understood without relying on the clarification traditionally provided in comments. Explaining why they had relegated the topic of documentation to the final chapter of the book, they wrote that documentation had "received proportionately too much attention—particularly in introductory courses—at the expense of clarity and general good style" (Kernighan and Plauger 1978, 6). In rejecting the necessity of commenting code in order to clarify the intentions of the programmer, *Elements* marked a schism between the priorities of most academic programming, including programming instruction, and what we now call software engineering, understood as a systematically managed, if highly remunerated profession and one offering significant autonomy. While those learning to program were (and still are) told to comment their programs liberally, both for their own benefit and for the convenience of their teachers, professional programmers producing and maintaining software applications for a living must manage the history of source code comments along with the code, continually reassessing them for relevance and accuracy in relation to changes in the code they were written to document.

"The form and approach of this book," Kernighan and Plauger wrote in their preface to the first edition, "has been strongly influenced by *The Elements of Style* by W. Strunk and E. B. White. We have tried to emulate their brevity by concentrating on the essential practical aspects of style" (Kernighan and Plauger 1978, xii). Where Strunk had introduced the 1920 first edition of *The Elements of Style* as addressed to "the principal requirements of plain English style" (phrasing that was adopted verbatim and retained in E. B. White's introductions to successive editions),[14] Kernighan and Plauger claimed for themselves a programming-linguistically agnostic or ecumenical approach, suggesting that "although details vary from language to language, *the principles of style are the same*. Branching around branches is confusing in any language. So even though you program in Cobol or Basic or assembly language or whatever, the guidelines you find here still apply" (Kernighan and Plauger 1978, 3, emphasis in original). Even as they invoked the cultural authority of their predecessors in formulating rules of style for written English in particular, Kernighan and Plauger thus insisted on generalizing their concept of style to any programming language then actively in use. Though this may be reasonable given not only the fact that all reserved words in programming languages were borrowed from English, but given also the symmetries observable in the designs of Fortran and PL/I (the languages of most of Kernighan and Plauger's examples), Algol, and to some extent Cobol (though the same could not be said of Lisp), this generalized insistence on style as a modality of program code in any language acted to further distance program code as ideally "self-documenting" from the natural-language domain of comments. "Code," Kernighan and Plauger wrote, "must largely document itself. If it cannot, rewrite the code rather than increase the supplementary documentation (Kernighan and Plauger 1978, 152).

Kernighan and Plauger presented their distaste for commenting in code as a generalized taste for succinctness of expression and concomitant efficiency in programmer labor. "Efficiency," they wrote,

> involves the reduction of overall cost—not just machine time over the life of the program, but also time spent by the programmer and by the users of the program. [. . .] Efficiency does not have to be sacrificed in the interest of writing readable code—rather, writing readable code is often the only way to ensure efficient programs that are also easy to maintain and modify. (Kernighan and Plauger 1978, 123)

In some of the cases they illustrated and analyzed at length, in *Elements*, comments that actively misrepresent the code they describe, either through direct error or through the passage of time producing "comment rot," are presented as more damaging than the absence of any comments at all. "A comment is of zero (or negative) value," they observed, "if it is wrong [. . .] The trouble with comments that do not accurately reflect the code is that they may well be believed subconsciously, so the code itself is not examined critically" (Kernighan and Plauger 1978, 141, 142). In Gerald M. Weinberg's *The Psychology of Computer Programming*, another classic early work of software engineering introspection whose first edition appeared three years before Kernighan and Plauger's book, the imagination of program comments as potentially dangerous distractions from (reading) code was developed into a theory of psychological "set" in computer programming and the specific distraction that comments presented to a programmer engaged in debugging, or searching for errors in a program, and the manager whose role it was to hustle him along and, ideally, to read over his shoulder. "Numerous experiments have confirmed," Weinberg wrote,

> that the eye has a tendency to see what it expects to see [. . .] Anyone who has ever tried proofreading is aware of this type of set phenomenon, and anyone who has ever tried to locate a mispunched word in a computer program is more than just aware—he is scarred [. . .] The whole idea of a comment is to prepare the mind of the reader for a proper interpretation of the instruction or statement to which it is appended. If the code to which the comment refers is correct, the comment could be useful in this way; but if it happens to be incorrect, the set which the comment lends will only make it less likely that the error will be detected. (Weinberg 1998, 162)

In a list of "alternative ways of looking at a program," Weinberg recommended that a programmer "strip comments from a listing and present the 'naked' listing on one page and the comments on another," segregating the two domains from each other entirely (Weinberg 1998, 266).

Weinberg described his work as "software psychology," but also as "software anthropology," and it is arguably the three chapters bearing the titles "The Programming Group," "The Programming Team," and "The Programming Project" that form the core of the book. Though the manager is a character here, described as an "outside threat" who works "for money or under threat" and finds it difficult to understand the programmer's attachment to intrinsically interesting work, it is also addressed to that character and written partly for his benefit. It was the manager's task, Weinberg suggested, to deal with "the problem of the ego in programming," training his programmers "to accept their humanity—their inability to function like a machine—and to value it and work with others so as to keep it under the kind of control needed if programming is to be successful" (Weinberg 1998, 56–57). As a domain of more or less complete syntactic and semantic freedom within the program, at least relative to the code itself, comments were one of several problematic sources of "programmer nonuniformity" (Weinberg 1998, 224) that an effective manager would have to contend with again and again. It would be fair to say that Weinberg's sympathetic but robust censure of the programmer, in his suggestion that program comments were ultimately more a hindrance than a help, articulates the managerial approach to what in any other context we would simply call *writing* as a threat to productive order.

By the early 1990s, the ideal of self-documenting code—that is, code containing no comments at all—exerted significant power in a now fully mature software engineering industry, one that had either put the software project management crisis of the 1960s and 1970s behind it or internalized that crisis at its core, depending on whom one asked.[15] In *Code Complete: A Practical Handbook of Software Construction*, first published in 1993 but still widely read today in its 2004 second edition, Steve McConnell declared that "the main contributor to code-level documentation isn't comments, but good programming" (McConnell 1993, 454). Programmers, McConnell argued, should strive for "the Holy Grail of legibility: self-documenting code. Such code relies on good programming style to carry the greater part of the documentation burden. In well-written code, comments are the icing on the readability cake" (Mc-Connell 1993, 456). Today, with the significant exception of the practice of so-called literate programming (discussed below), the ideal of self-documenting code is, if still seldom achieved, nonetheless a point of consensus. A more recent entry into the market for manuals of software craftspersonship, Pete Goodliffe's *Code Craft: The Practice of Writing Excellent Code* (2007), bluntly advises working programmers to "always have faith in code and doubt comments" (Goodliffe 2007, 86).

At the same time, and despite the excommunication of the natural-language domain of comments accomplished by Goodliffe's advice, as well as his concept of source code as self-documenting document ("Your source code is a document, describing the program you are creating [. . .] The only document that describes your code completely and correctly is the code itself"),[16] the emphasis Goodliffe places on readability leads him to argue from analogies to the natural-language domain of reading and writing, analogies that arguably confuse the segregation of code more than they clarify it. Indeed, insofar as such analogies are foundational to the practical discourse on programming style that emerged with Kernighan and Plauger's *Elements*, we might understand such vacillation as an inheritance of the language metaphor through which computer programming in particular (rather than data processing more generally) was conceptualized and described, in the early history of computing.[17]

When, in an essay titled "Programmer as Reader," Adele Goldberg declared that in programming practice, "reading and writing are part-ners," that "readability is an issue because we read to learn to write," and that "readability issues are strongly tied to the ability to write" (Goldberg 1987, 62), she was drawing on this long history of the language metaphor in computing in the most general sense. But the history of programming languages and their environments, in the broadest sense of that word, provides us with a larger context for an apparent consensus about self-documenting code.[18] "Unlike the paper-and-pencil approach to writing a novel that will be distributed as a hard-copy book," Goldberg wrote, "the programmer's writing is available within a dynamic system" (Goldberg 1987, 70). Although such a distinction might well apply to *any* use of a programming language, what Goldberg had in mind was a "newer ex-ploratory programming style of development" (Goldberg 1987, 62) made possible by integrated environments such as the Smalltalk-80 system that Goldberg had designed with Alan Kay and Dan Ingalls from 1969 to 1972. Smalltalk's deep and wide-ranging influence on the history of programming language design, an influence that extends all the way to systems programming, must be set alongside its origins in instructional computing, specifically for children, and a characteristic feature that reflects that origin in design for instructional and expressive computing. Unlike most other programming languages as such, Smalltalk is not a programming language in the usual sense at all, but an integrated runtime environment in which program code has no meaningfully static form of the kind that industrial software craftspersonship, with its focus on readably "clean" code, requires as an object in order to present itself as professional discipline. That is to say that Smalltalk program code is not a document in the usual sense, or even in the extended sense in which program text has the special feature of executability. Instead it represents, for lack of a better word, merely the state of a runtime environment at a particular moment.

In this very specific, if not at all incidental aspect of its design, Smalltalk-80 is unquestionably unusual, even exceptional in the history of programming, notwithstanding the widespread adoption of many of its syntactic features into other programming languages proper and its early elaboration of the object-oriented programming paradigm that would come to dominate both academic instruction and professional practice for almost three decades beginning in the 1980s. Although Smalltalk variants have a typographically demarcated comment syntax like any other, their integration of static program code (as text, as document) into an execution environment running on a virtual machine also has consequences for the domain relations between program code and comments. The activity of programming in Smalltalk is not, as with most other language implementations, an activity occurring in defined steps or phases, in which writing code is followed by the mutually distinguishable program life cycles of compilation and execution, then the rereading and rewriting of code and its compilation and execution again. The composition of program comments to preserve reflection on aspects of this process presumes that a program is a static document developed in phases of reading and writing interleaved with compilation and execution (including testing); but the programmer writing ideally self-documenting code, understood as a subsumption or displacement of comments, presumes so as well. The user of a Smalltalk system, by contrast, is effectively programming that system as it runs in real time, with the results of execution integrated into the loop in which the reading and writing of both code *and* execution results or effects appear to converge in the presentation of the system as "live": a living document, if we must put it that way in deference to the way programming is usually understood (that is, as a dynamic operation on versions of a static document).[19]

But again, all this sits outside the consensus of a fully mature software engineering industry, in which the *machine* readability of code is paramount and the purpose of the writing of code is to get it to run, not to reflect on the process of writing in Goldberg's generous and humanistic sense. The ideal of commentless or self-documenting code, purified of all comments whose very existence testifies to and continues to request human reading, which can be as abstruse as the code they describe, requesting still more reading, and which are all too often orphaned as a codebase evolves, becoming sources of confusion and inconsistency: all this aligns with the fundamental technical logic of software as automation. The mandate to write self-documenting code reduces distractions for the human programmer, eliminating the bicameralism of writing code to perform a task and comments to explain why it performs the task in that particular way, rather than any other. Self-documenting code is a model for automatic code generation in any of many possible modes, from the basic use of templates or other forms of preexisting code, up to the wholesale production of working code by sophisticated utilities like OpenAI's Codex. A human programmer writing perfectly self-documenting code is actively anticipating automatic code generation, in fact is *becoming* an automatic code generator, at least as an ideal. All of this is aligned with the core purpose and exclusive function of software, intrinsically speaking, which is to automate other tasks. Comments have always represented speed bumps on the way to automation.

Literate Programming

Whereas self-documenting static program code has been purified of equally static program comments, so-called literate programming very nearly does the reverse, consuming program code in a profusion of "internal" documentation that turns a computer program-as-document as it were inside out, making its documentation an enveloping shell. We might say that literate programming represents a hypostatization of the incommensurability of program code with comments that is suggested

by Michael Van De Vanter's argument that "the documentary structure
of source code is not related to the linguistic structure in any tractable
way" (Van De Vanter 2002, 773) and that "the documentary structure of
code is grounded in information that *cannot be derived from* its linguistic structure, and in fact *cannot even be understood* in those terms" (Van
De Vanter 2002, 768; original emphasis).

In July 1968, Charles H. Lindsey published in *ALGOL Bulletin* a forty-
one-page typescript essay that began with what one might at first take for
document headers or section titles, albeit oddly formatted ones. The first,
indented eight spaces from the left margin, was the underscored, uncap-
italized word "**begin**". The second, beginning flush at the left margin,
read "**comment 1. INTRODUCTION**." Following these two lines of text,
there appeared this sentence, indented eight spaces: "This is a program
written in ALGOL 68" (Lindsey 1968, 9).

"ALGOL 68 with Fewer Tears" was an essay composed for readers of
the English-language *ALGOL Bulletin*, a periodical published beginning
in 1959 that became a record for the activities of the International Federation for Information Processing (IFIP) Working Group on Algorithmic
Languages and Calculi and which in February 1968 had published the
working group's draft report for Algol 68, a significant revision of Algol
60. The vast majority of the text filling its forty-one pages was conven-

AB*28* p *9*

ALGOL 68 WITH FEWER TEARS   C.H.Lindsey.
(2nd edition)

    begin
comment 1, INTRODUCTION
        This is a program written in ALGOL 68. It is (or at least is intended
to be) a valid program, syntactically, but it does not purport to do anything
sensible. It comprises my interpretation of what the 'Report' is all about,
the 'Report' being the
           Draft Report on the Algorithmic Language ALGOL 68.
           Edited by A. van Wijngaarden, and published by
           the Mathematisch Centrum, Amsterdam.

FIGURE 4. Image of first page of C. H. Lindsey, "ALGOL 68 with Fewer Tears"
(1968). Reprinted with the permission of International Federation for Information Processing.

tional technical English prose in a mode suiting its subject, occasion, and venue, a technical bulletin. "ALGOL 68 with Fewer Tears" was also, however ("or at least is intended to be," as Lindsey put it) a valid Algol 68 program, though one that did "not purport to do anything sensible." "It comprises," Lindsey continued, "my interpretation of what the 'Report' is all about" (Lindsey 1968, 9).

Following two short paragraphs in which Lindsay commented on the relationship between Algol 60 and Algol 68 and reminded his readers that he was not himself a member of the working group that produced the latter, only an interpreter of the published draft report, Lindsay began explaining the syntax of Algol 68. "A program in ALGOL 68," he began, "consists of a block enclosed between **begin** and **end**. This is itself presumed to be enclosed within an outer block within which a variety of built-in procedures, etc. and library routines are declared" (Lindsey 1968, 9). "Comments," he continued,

> may be inserted anywhere within a program (i.e. not just between statements), except in the middle of compound characters such as **begin** and **end**. I am in the middle of a comment at the moment, and I shall terminate it by writing **comment**
>
>   **comment** is also used at the beginning of a comment (we are in another one now). As an alternative to comment **c** may be used, at either end. These two words, of course, cannot be used within comments. (Lindsey 1968, 9)

Most of the forty-one pages of text that made up "ALGOL 68 with Fewer Tears," it should now be clear, consisted of comments inserted into a short Algol 68 program that was at least theoretically valid and executable in conformance with the working group's draft specification for the language. (Since no Algol 68 compilers were yet available, this could not be tested.) To put it in generic terms, rather than reading a forty-one-page essay providing examples and illustrations of Algol 68 syntax as specified in the working group's draft report, what Lindsey's readers read in reading "ALGOL 68 with Fewer Tears" was precisely the reverse: an Algol 68 program whose dependent text or subdocument was an essay explaining Algol 68 syntax. In theory, at least, a reader of the essay could also compile it and observe its results as a program.

With minimal changes, "ALGOL 68 with Fewer Tears" was republished in typeset form in the *Computer Journal* in 1972 with a headnote remarking that "many implementations of ALGOL 68 are under way, and the one implementation actually running has successfully compiled this paper" (Lindsey 1972, 176).[20] It was to this version that Donald E. Knuth was referring when he wrote twelve years later, in the same journal, that "the fact that at least one paper has been written that is a syntactically correct ALGOL 68 program encourages me to persevere in my hopes for the future. Perhaps we will even one day find Pulitzer prizes awarded to computer programs" (D. E. Knuth 1984, 112). It is worth quoting at length from the first three paragraphs of Knuth's essay, titled "Literate Programming":

> The past ten years have witnessed substantial improvements in programming methodology. This advance, carried out under the banner of "structured programming," has led to programs that are more reliable and easier to comprehend; yet the results are not entirely satisfactory. My purpose in the present paper is to propose another motto that may be appropriate for the next decade, as we attempt to make further progress in the state of the art. I believe that the time is ripe for significantly better documentation of programs, and that we can best achieve this by considering programs to be *works of literature.* Hence, my title: "Literate Programming."
>
> Let us change our traditional attitude toward the construction of programs. Instead of imagining that our main task is to instruct a *computer* what to do, let us concentrate rather on explaining to *human beings* what we want a computer to do.
>
> The practitioner of literate programming can be regarded as an essayist, whose main concern is with exposition and excellence of style. (D. E. Knuth 1984)

"Literate Programming" would become a 1992 book with the same title, as well as a paradigm (or movement) of some consequence in academic computer science and perhaps academic computing slightly more generally, if nowhere else. In "Computer Programming as an Art," Knuth's 1974 A. M. Turing Award lecture (also reprinted in the volume *Literate Programming*), Knuth had approvingly referred his audience to Kernighan and Plauger's *The Elements of Programming Style*, as evidence of welcome attention to the idea of style in programming;[21] but Knuth's concept of literate programming was in many ways diametrically opposed to the commentless "clean code" approach proposed by *Elements.* The former emerged from Knuth's experience using a document preparation system of his own design, and was tightly coupled to that system, which Knuth had named WEB.[22] (Knuth tells us that he had chosen this name "partly because it was one of the few three-letter words of English that hadn't already been applied to computers" [Donald E. Knuth 1992b, 100].) WEB was a specification integrating a document typesetting language with a programming language; in Knuth's earliest personal implementations the former was TeX, the document typesetting language that Knuth had been developing since 1977, and the latter was Pascal (Donald E. Knuth 1992b, 101). "A WEB user," Knuth wrote, "writes a program that serves as the source language for two different system routines." In processing that program document, a single input artifact, the WEB system produced two quite separate output artifacts: one a typeset document suitable for printing and reading as an essay, and the other a compiled executable object.

One advantage offered the programmer by such a system was that "the program and its documentation are both generated from the same source, so they are consistent with each other" (Donald E. Knuth 1992b, 101). There was no longer any need to manually update a program's external documentation each time the program was modified, to ensure that the documentation accurately reflected that modification. More significant was the assistance that Knuth argued such a system provided for incremental and essayistic work on a program and its capacity to preserve, for the reader of a literate program, the original order of its components' creation. In creating the second of its two output artifacts, the compiled executable object, WEB could assemble (Knuth's word was "tangle") the components of a working program in a different order than the order in which they appeared in the program, thus relieving the programmer of the obligation to reorganize those components, in the process obscuring or destroying the program's original form. Knuth argued that both writer and reader of a program stood to benefit from this preservation of the record. "My experiences have led me to believe," Knuth wrote,

> that a person reading a program is, likewise, ready to comprehend it by learning its various parts in approximately the order in which it was written [. . .] the WEB language allows a person to express programs in a 'stream of consciousness' order. TANGLE is able to scramble everything up into the arrangement that a Pascal compiler demands. This feature of WEB is perhaps its greatest asset; it makes a WEB-written program much more readable than the same program writ-

ten purely in Pascal, even if the latter program is well commented. (Donald E. Knuth 1992b, 125–26)

This latter contrast is decisive: as Douglas McIlroy put it in an assessment of one of Knuth's demonstration literate programs, "in WEB one deliberately writes a paper, not just comments, along with code."[23] Other commentators, no doubt recalling Knuth's own emphasis on narrative, have reached for different generic designators: "A literate program," Goodliffe suggests, "is written almost as a story; it is easy for the human reader to follow, perhaps even enjoyable to read. It is not ordered or constrained for a language parser. This is more than just a language with inverted comments; it's an inverted method for programming" (Goodliffe 2007, 66). We might say that in literate programming, comments are *extruded within* a program, rather than being embedded in it as what Van De Vanter called "miniature natural language documents" (Van De Vanter 2002, 776), and that this extrusion extends the documentary structure of source code to its limit, in ways that also touch questions of genre that have yet to be explored. Indeed, Van De Vanter suggests that literate programming represents "the most ambitious and successful attempt to rethink the relationship between documentary structure and source code," despite having "never been widely adopted" (Van De Vanter 2002, 779). This is not even to mention other, more explicitly political and political-economic analogies through which computer programming has been reimagined as literary composition, such as those explored by Richard P. Gabriel and Ron Goldman in their manifesto for "mob software" as code-as-literature-opposing-capital:

> Fast-lane capitalism has created a nightmare scenario in which it is literally impossible to teach and develop extraordinary software designers, architects, and builders. The effect of ownership imperatives has caused there to be no body of software as literature. It is as if all writers had their own private companies and only people in the Melville company could read "Moby-Dick" and only those in Hemingway's could read "The Sun Also Rises." [. . .] When software became merchandise, the opportunity vanished of teaching software development as a craft and as artistry. The literature became frozen [. . .] Mob software takes coding out of the closet and makes code literature. (Gabriel and Goldman 2000)

Though we can safely place beyond the pale Gabriel and Goldman's wishful vision of "the literature of code" as "a creativity heretofore rarely observed in software development," one wonders if Van De Vanter's dismissal of literate programming as a niche pursuit may have been premature given the profusion of tools and environments inspired by Knuth's ideas that emerged in the later 2000s and their growing popularity not only within the academic sciences but in so-called "data science" industries and in newly data-intensive modes of journalism as well.[24]

Conclusion: "R. I. P. L. V. B."

Program comments have long been put to ends that suit the paradigms of neither clean-code software craftspersonship nor literate programming. The recording of program metadata or authorship information, assertions of intellectual property or other legal boilerplate, corporate logos, even diagrams and other forms of technical illustration all represent less common but enduring professional, rather than idiosyncratic uses of the documentary structure of source code. Possibly the most interesting dimension of this interstitial (or leaky) domain of professional activity is its moonlit side, the idiosyncratic use and abuse of program comments that record traces of programmer labor.

Handbooks of software craftspersonship are unanimous in their denunciation of idiosyncrasy in program comments. "Little witty comments may be witty, and they may be little," opines Goodliffe, "but *don't* put them in. They get in the way and cause confusion. Avoid expletives, inside jokes that only you understand, and comments that are unnecessarily critical—you never know where your code will end up in a month or year's time, so don't write comments that could cause you embarrassment later" (Goodliffe 2007, 77). While profanity in comments in the source code of the Linux kernel (just for example) is a source of amusement,[25] it is also regarded as a symptom of inconsistent discipline in both academic and so-called FOSS (free and open-source software) programming practice, along with the insertion of verse (limericks and haiku are favorites), ASCII art banners,[26] and undocumented "Easter egg" features. McConnell supported his advice to "avoid self-indulgent comments" with the following anecdote:

Many years ago, I heard the story of a maintenance programmer who was called out of bed to fix a malfunctioning program. The program's author had left the company and couldn't be reached. The maintenance programmer hadn't worked on the program before, and after examining the documentation carefully, he found only one comment. It looked like this:

MOV AX, 723h   ; R. I. P. L. V. B.

After working with the program through the night and puzzling over the comment, the programmer made a successful patch and went home to bed. Months later, he met the program's author at a conference and found out that the comment stood for "Rest in peace, Ludwig van Beethoven." Beethoven died in 1827 (decimal), which is 723 (hexadecimal). The fact that 723h was needed in that spot had nothing to do with the comment. Aaarrrrghhhhh! (McConnell 1993, 468).

With tongue in cheek, to be sure, we might call such examples "literary programming," to distinguish them clearly from the purposiveness of Knuth's practice. Other examples might include the source code for the garbage collection routines of the MIT implementation of the Lisp dialect Scheme, whose documentation comment begins with the sentence "Open wabbit season on wabbits matching WABBIT-DESCWIPTOR and go wabbit hunting" and maintains its idiom throughout, including in the naming of functions and other expression elements,[27] or even the program code for the flight software run by the Apollo 11 program's Apollo Guidance Computer. The latter code had been publicly available since 2003, initially in scanned pages of printouts and eventually in transcription, but in July 2016, a former US National Aeronautics and Space Administration intern made the transcriptions available on GitHub, a web-based hosting service for code and document repositories, which brought them to wide public attention. In the annotations of assembly language instructions for guiding a spacecraft, one observes a perhaps peculiar levity: "HELLO THERE" followed by "GOODBYE. COME AGAIN SOON"; "OFF TO SEE THE WIZARD"; "ASTRONAUT: PLEASE CRANK THE SILLY THING AROUND"; "TEMPORARY, I HOPE HOPE HOPE" (indicating that the programmer never intended their decision to be final; yet it was). There are also literary excerpts and allusions, such as the verse from 2 Henry VI which appears to comment on the syntax of the Apollo Guidance Computer instruction set, and the injunction HONI SOIT QUI MAL Y PENSE.[28]

Some of the types of comments that McConnell rejected as "self-indulgent" are implicitly managerial expressions of frustration with code written by other programmers, or anticipations of the frustration of readers of one's own code; others arguably serve as cleaner windows onto the realities of programming as thoroughly managed labor, precarious or abject in its own ways and very much a target for automation. Peruse any of the many available compilations of sourced, anecdotal, quoted, and embellished "funny" or "witty" comments, and you will find more or less entertaining iterations of exclamations and interjections like "huh?," "eh," "wtf?," "yikes," "scary," "here be dragons," "shoot me now," "brutal hack," "please work," "magic, do not touch," "drunk, fix later," and "abandon all hope, ye who enter here"—but also "sorry for my laziness," "there are dumber fucks than me out there, but not many," and "I'm sorry. —Old programmer who was about to retire with an ex-wife and two kids."[29] "Dear future me. Please forgive me. I can't even begin to express how sorry I am" and "When I wrote this, only God and I understood what I was doing; now, God only knows" are autobiographical and auto-managerial, comments that might be read as lamenting either the absence of comments or the failure to write commentless clean code. "If you're reading this, that means you have been put in charge of my previous project. I am so, so sorry for you. God speed" is an apology directed to other programmers, while "fucking imap fucking sucks. what the FUCK kind of committee of dunces designed this shit" and "Having worked on this code for several weeks now, my hate for PSD has grown to a raging fire that burns with the fierce passion of a million suns" indict them.[30] "You can't fix this, but please increment the counter below if you try [. . .] hours wasted here: 56" foregoes rage for the (manual) tabulation of such struggle.[31]

Casual as they are, comments that dwell explicitly on programmers' working conditions and the realities of both enterprise and startup software industries might be valued as documenting labor conditions that in many ways, as I suggest in this book's introduction, have yet to be adequately theorized in their specificity. "I have to find a better job," "This isn't the right way to deal with this, but today is my last day," "uncomment the following line if the program manager changes her mind again this week," "I am not responsible of [sic] this code. They made me write it, against my will," and "I dedicate all this code, all my work, to my wife, Darlene, who will have to support me and our three children and the dog once it gets released into the public"[32] use the documentary structure of source code to ends that manuals of software craftspersonship strive explicitly to discipline and that literate programming, a predominantly academic pastime, has no purpose for. Here, too, at another level, one more directly integrated with the domain of the social generally, what I have been calling the apposition and adjacency of code with comments in a program evinces a specific character.

It would not be too much to suggest that in its hostility to comments, the discipline of software craftspersonship has, programmatically in some cases and perhaps blindly in others, been constructed to extinguish this side channel of communication, whose apposition protects it from interfering with the programmer's work (that is, with writing a working program) but whose adjacency documents the programmer's work in ways that undermine, often profoundly, its imagination as fun work or creative labor[33]—an imagination on which both programmers themselves and their managerial cadres, in the so-called technology industry in the United States in particular (but not only there), are profoundly reliant for their prospects and privileges. That this imagination, those prospects and privileges, have confined the prospects of others, with direct or indirect but in any case unwanted, indeed extremely unpleasant consequences is, perhaps, now beginning to be grasped, albeit perhaps too late. Could we do worse than to crash the program by *really* reading the code?

––––––

Focused on a contact zone within a computer program where natural languages and programming languages appear to touch or intermingle, this chapter has continued the discussion of automation begun in chapter 1, suggesting that computer programming can be understood both as a kind of writing and as a special class of human labor activity facilitating the automation of other human labor activities, including itself. As we have seen, comments represent a space of relative freedom within the constraint of a program understood as the automation of automation, and programmers have used comments to document not only the code of a program but also the activity and the experience of programming as a self-automating job and profession. In the following chapter, focused on so-called metasyntactic variables, we will examine a closely related topic: the freedom to choose names and its legislation in programming language style guides, manuals of software craftspersonship, and management theory focused on cultures of software development.

THREE

Etymologies of Foo

In chapter 2 I presented a study of the program comment, treating it as a contact zone within a computer program where natural languages and programming languages appear to touch or intermingle. Program comments are not, of course, the only such area of interest to the philological approach I take in this book. This chapter focuses on another such contact zone, in variable naming in general and in a class of natural-language artifacts called metasyntactic variables, in particular.[1]

"Anyone could say anything and nothing was official"

On April 1, 2001, Network Working Group RFC (Request for Comments) 3092, authored by Donald Eastlake III of Motorola Inc., Carl-Uno Manros of the Xerox Corporation, and Eric S. Raymond of the Open Source Initiative (OSI), was published with the title "Etymology of 'Foo'." "Approximately 212 RFCs so far," the document's abstract observed, "contain the terms 'foo', 'bar', or 'foobar' as metasyntactic variables without any proper explanation or definition. This document rectifies that deficiency" (Eastlake, Manros, and Raymond 2001, 1).

The philological promise of this document's title might be described as at least partly tongue in cheek, in so far as its occasion was ad hoc and its justification was as revisionist as it was inceptive. As a Request

for Comments, "Etymology of 'Foo'" was added to the technical document series by that name (more properly, by the name Internet Requests
for Comments) that had been launched in 1968 along with the first US
Advanced Research Projects Agency (ARPA) grant supporting the development of a protocol for linking computers across a telecommunications network. In an arrangement that might be considered idiosyncratic
today, ARPA's funding supported a loosely coordinated team of researchers composed mostly of graduate students at the Los Angeles and Santa
Barbara campuses of the University of California, the University of Utah,
and MIT, who conferred with representatives from the RAND Corporation and Bolt, Beranek & Newman, the high-technology research and services firm headquartered in Cambridge, Massachusetts. The procedural
informality of the Network Working Group (NWG) that began meeting
in 1968 reflects the unusual circumstances of the moment, which historians of the early internet have attributed to factors including an early
(but typical) focus on hardware, which left the ostensibly less important
work on software to students, and to the team's ad hoc formation, distributed location, and closure as a research community.[2] By all accounts,
the founding student members of the NWG were surprised to find themselves left to their own devices: Stephen D. Crocker would later reflect
that "most of us were graduate students and we expected that a professional crew would show up eventually to take over the problems we were
dealing with." When the Bolt, Beranek & Newman team arrived for their
first meeting with the fledgling NWG, they "found themselves talking to
a crew of graduate students they hadn't anticipated [. . .] while BBN didn't
take over the protocol design process, we kept expecting that an official
protocol design team would announce itself" (Reynolds and Postel 1987,
2). That never happened: as Andrew L. Russell has put it, "Crocker slowly
realized that no such team was going to arrive. Much to his surprise, the
team of graduate students and contractors in NWG had uncontested ju-
risdiction over protocol development for the Arpanet" (Russell 2014, 169).

What is interesting about the early history of the NWG is the process
by which its improvised expertise was deliberately formalized as informal communication and record-keeping. "After a particularly delightful
meeting in Utah," Crocker recalled in the same memoir,

> it became clear to us that we had better start writing down our dis-
> cussions. We had accumulated a few notes on the design of DEL[3] and

other matters, and we decided to put them together in a set of notes. I remember having great fear that we would offend whomever the official protocol designers were, and I spent a sleepless night composing humble words for our notes. The basic ground rules were that anyone could say anything and that nothing was official. And to emphasize the point, I labeled the notes "Request for Comments." (Reynolds and Postel 1987, 2–3)

As the story of how the RFCs became "the vehicle for the Network Working Group to publish consensus statements and technical standards for the Arpanet—even though they were specifically intended *not* to become standards" (Russell 2014, 169) is now fairly well known, I will not review it here, though its less well-documented implications for a genre theory of technical communication might also be noted. Janet Abbate reflected that while the RFCs were initially printed on paper, they were stored in electronic form and accessed via the Arpanet as soon as it was possible to do so, and that their role in evolving "formal standards informally" cannot be delinked from their then-peculiar mode of production, distribution, and access, notwithstanding Crocker's statement that he had "never dreamed these notes would [be] distributed through the very medium we were discussing in these notes."[4]

The informality of the process by which RFCs were first published, preserved, and used for reference ensured that the documents retained traces of the deliberative process that would normally be removed from the record, or which the creation of a record would normally suppress altogether. Sandra Braman notes that for some time "there appears to have been no expectation that anyone outside of that community would ever be looking at the documents" (Braman 2011, 298)—one of several possible reasons that anger, humor, speculative metacommentary on the deliberative process or its direct narration, and other such social byproducts of group decision-making were retained in early RFCs. Though Brian E. Carpenter and Craig Partridge's claim for RFCs as scholarly publications may rest on an idealized understanding of "open" review and a deliberately enervating contrast with other "scholarly publication systems," they are right to emphasize the effects of a later, conventional formalization of the RFCs as the collection's purpose evolved from exploratory communication to authoritative publication, during the later 1970s (Carpenter and Partridge 2010, 31).

RFC 3092 "Etymology of 'Foo'" appeared long after the document

series embraced this more conventional, authoritative mode. In its reflec-
tiveness on the past and its stated intention to redress some specific, now
less desirable consequences of the in-group informality of that past, RFC
3092 reflects the outward-directed perspective of the RFC as a mature
form—that is, as a publication in the conventional sense, directed to
some reading public not all of whose members were necessarily known
in advance. To a limited yet genuine extent, its purpose, to provide an
etymology of the English-language word "foo," can be described as scholarly. And yet insofar as its topic is a philological topic, quite clearly more
cultural and historical than technical in nature, RFC 3092 inhabits an
eccentric minor subcategory of RFCs, perhaps closer to those in category
1G "Request for Comments Administrative," as catalogued in RFC 1000
"Request for Comments Reference Guide,"[5] than to others now carrying
the standardized "Informational" designation assigned to many RFCs
that earn that descriptor, but also to the parodic RFCs often published
on April Fool's Day each year. In this and in other ways that make it
neither an essential technical document, nor one clearly segregatable
with openly or hyperbolically humorous and "joke" RFCs, "Etymology of
'Foo'" might be understood as representing the substantive, rather than
glib or flippant persistence of informality within the formality of the RFC
series of today. I suggest that the putative topic of "Etymology of 'Foo',"
the metasyntactic or deliberately meaningless word given in its title, is as
good a marker of this persistence as any; and that in like manner, in their
specific contexts in program code in so-called higher-level programming
languages, the term "foo" and associated terms are specifically meta-
abstractive, abstracting the abstraction of higher-level languages, which
is to say rendering their abstraction concrete and tangible.

Reserved and Unreserved Words

"Approximately 212 RFCs, or about 7% of RFCs issued so far, starting
with RFC 269," Eastlake, Manros, and Raymond began,

> contain the terms "foo", "bar", or "foobar" used as a metasyntactic
> variable without any proper explanation or definition. This may seem
> trivial, but a number of newcomers, especially if English is not their
> native language, have had problems in understanding the origin of
> those terms. (Eastlake, Manros, and Raymond 2001, 1)

RFC 3092 "Etymology of 'Foo'" concluded with an appendix containing a table of occurrences of the words "foo," "bar," "foobar," and "fubar" in RFCs 269 through 3092 itself. In the document's five intervening pages Eastlake, Manros, and Raymond provided brief definitions of the words "bar," "foo," "foobar," and "foo-fighter," in that order, drawing disproportionately on Raymond's own *The Jargon File*[6] but also on the *Oxford English Dictionary*, the *Free On-Line Dictionary of Computing*, and Ron Goulart's *Encyclopedia of American Comics*, among other sources. Only two of the six definitions given for "foo" defined the word as it had been used in existing RFCs to date. The first of these two, brief and reproduced in its entirety from *The Jargon File*, was "Used very generally as a sample name for absolutely anything, esp. programs and files (esp. scratch files)" (Eastlake, Manros, and Raymond 2001, 1).

The contemporary English word "foo," for which the *Oxford English Dictionary* provides no definition, appears to date to the 1930s. It can be found in editions of *The Tech*, the student newspaper at MIT, as early as 1937, in usage contexts that remain unclear.[7] It may have originated in Bill Holman's comic strip *Smokey Stover,* created in 1935, where it appeared as a nonsense word and in mostly nonsensical catchphrases like "foo fighter," "foomobile," "firefighter of foo," "fire chief of foo," and "where there's foo, there's fire."[8] The migration of "foo fighter" into US Air Force jargon during World War II[9] has encouraged the imagination of a genealogical relationship to other US military slang terms of the same period, including the acronyms "snafu" ("situation normal, all fouled/fucked up") and "fubar" ("fouled/fucked up beyond all recognition").[10] Students and others in MIT's Tech Model Railroad Club, formed immediately after the war, included it in their club "dictionary" with a jocular definition consistent with its cryptic usage in *The Tech*.[11]

To all appearances, it was not until the early 1960s that the latter usage, which conforms to the first part of Eastlake, Manros, and Raymond's initial definition ("a sample name for absolutely anything"), was extended into that definition's second part ("esp. programs and files"). Today the only speakers and writers of US English who can be counted on to recognize the word "foo" are academic computer scientists and professional software developers, who will also almost certainly also know much of the "standard list of metasyntactic variables" provided in the second of Eastlake, Manros, and Raymond's two RFC-specific definitions. "Foo" and "bar," along with "baz," "qux," and "quux," in that order, followed less strictly by "corge," "grault," "garply," "waldo," "fred," "plugh," "xyzzy," and "thud," among others, are used as placeholder names in program code that may be sketched quickly and may be intended to be provisional, demonstrative, illustrative, or otherwise temporary. *As* a specification, the specification of a programming language (that is, its formal definition in a document intended for use as a guide to implementation) is intrinsically constrained: a basic and obvious characteristic that is nonetheless frequently elided by excitable analogies binding code to language, including the concept of a "programming language" itself.[12]

As a designed system, a programming language is determined by constraints incommensurable with the arbitrary historical finitude of a natural language. Only its reserved words or identifiers—a small number of syntactic objects given precise semantics—are lexed and parsed, or compiled, usually in the form of blocks of instructions (including functions, classes, and methods): that is, as code to be run. Other text in a program represents data, is ignored (for example, program comments, which are demarcated using a border or barrier symbol), or violates rules of syntax, producing compiler error. Finally, there are identifiers chosen by the program writer as names for classes, functions or methods, and constants or variables. Provided that they are created correctly, according to specified (and rigid) orthographic rules, and that they are not identical to reserved words or identifiers, such names might be words in any natural language whose natural-language meaning, as presented by a dictionary, clarifies a program segment's intent, purpose, or structure. A programmer writing a program to compute a financial transaction might, for example, choose the names *price, tax, shipping,* and *total* for variables assigned the relevant values at stages of the transaction; such a programmer might give a compound name like *calculateTax* or *calculateShipping* to functions or methods that perform subtasks supporting the final result.

It may seem dramatic to suggest that in a domain of such radical constraint, this freedom to choose names becomes a site of struggle, even a kind of suffering, as the programmer is faced with elaborating ad hoc a second, entirely separate, exclusively human-readable task- and program-specific semantics interlayered with that of strictly determined machine-readable syntax, in those portions of the program that run. But if this freedom were not also a curse, it would not be the object of so much legislation in programming language style guides and manuals of software craftspersonship.[13] In practice, such conventions tend to reflect the domain of use, from the single-letter names ("a," "b," "x," "y") common in academic computer science, which reflect the latter's kinship with mathematics, to the very long compound names ("InternalFrameInternalFrameTitlePaneInternalFrameTitlePaneMaximizeButtonWindowNotFocusedState") found in a large enterprise software project.[14] But controversy and conflict over naming conventions can be found even within the most orthodox and best regulated domains of use.

The choice of "foo," "bar," and so on as names thus has two quite distinct contexts. In one context, it represents a deliberate choice of the asemantic abstraction of a so-called nonsense word—as is common in computer science research and instruction, or in the sample code provided in software library or API (Application Programming Interface) documentation, where a high level of abstraction follows the requirement to provide generalizable examples. Relieved of ready or apparent reference, such tokens stand in for domain- or application-specific signifiers whose specification would be premature in such contexts, obscuring the pedagogical generality and availability of the example. In principle, the data that "foo" or "bar" makes available to some later procedure might represent anything from the integer value 1 to the memory address of an intricate data structure.

In another context, however, the choice of "foo" or "bar" represents a deferral of the choice of a semantically (more) meaningful word (for example, "calculateTax," "calculateShipping," or "InternalFrameInternalFrameTitlePaneInternalFrameTitlePaneMaximizeButtonWindowNotFocusedState"), where the latter choice is necessary and required eventually. The latter is the context of professional software development, where the primary emphasis is on tasks and purposes rather than concepts and where both software creation and software maintenance are performed by teams, accounting for the premium placed on code standards including standard styles and other measures of readability. Here, too, "foo" or "bar" holds a place for later specification, but that specification is anything but premature; indeed, it serves a designated domain and design. The only thing standing in its way is the temporality and the sustainability of the programmer's labor in their culmination in a readable, maintainable, tested and documented internal or external release of code.

In both of these technical usages as a metasyntactic variable in program code, the word "foo" can be found in programming textbooks and other instructional and noninstructional documentation from the mid-1960s onward.[15] "Foo" and "bar" appear both separately and in conjoined form, though invariably in sequence (a characteristic encouraging the imagined link to the aforementioned acronym "fubar"). John Everett recalls that when he began working at Digital Equipment Corporation in 1966,

> foobar was already being commonly used as a throw-away file name. [. . .] Foo and bar were also commonly used as file extensions. Since the text editors of the day operated on an input file and produced an output file, it was common to edit from a .foo file to a .bar file, and back again. It was also common to use foo to fill a buffer when editing. (Everett 1996)

Of the examples of metasyntactic variables provided in *The Jargon File*, most of which date to the 1970s, Raymond noted that "the word *foo* is the canonical example. To avoid confusion, hackers never (well, hardly ever) use 'foo' or other words like it as permanent names for anything. In filenames, a common convention is that any filename beginning with a metasyntactic-variable name is a scratch file that may be deleted at any time" (Raymond 2004a).

Raymond catalogued variations on the series, adding notes on their supposed provenance, such as "foo, bar, thud, grunt" ("This series was popular at CMU [Carnegie Mellon University]"); "foo, bar, bletch" ("Waterloo University. We are informed that the CS [Computer Science] club at Waterloo formerly had a sign on its door reading 'Ye Olde Foo Bar and Grill' "); "foo, bar, fum" ("This series is reported to be common at XEROX PARC [Palo Alto Research Center]"); "foo, bar, baz, bongo" ("Yale, late 1970s"); and "foo, bar, zot" ("Helsinki University of Technology, Finland"). Noting also ostensibly British ("fred, jim, sheila, barney"), Dutch ("aap, noot, mies"), French ("toto, titi, tata, tutu"), Italian ("pippo, pluto, paperino"), and New Zealand ("blarg, wibble") variations, among others, Raymond suggested that "of all these, only foo and bar are universal" (Raymond 2004a).

Name, Variable

In both programming language design and programming practice, the word "name" is used for the human-readable linguistic token chosen by the programmer and stored in an encoded form including a location address in system memory, where data associated with the name will be stored. A programmer chooses the name, but not the memory address. The latter is assigned by the programmer's software tool called an interpreter, compiler, or assembler, which associates the name with the address in a process or event called *binding*.

"Variable" is one of several possible terms for the composite computational object formed by binding a programmer-chosen, human-readable name to a memory address holding associated data. In many programming languages, almost any association between a name and a memory address of stored data, from a single binary value to an entire software package or namespace, can be described as a binding: here, the distinction that matters is between a binding that can be changed (that is, which is mutable in character) and a binding intended to be constant, which cannot be changed. Used more strictly, "variable" refers to (mutable) bindings to discrete, elementary data objects—usually noncompound objects, often the smallest and least important to the general purpose of the program, explicitly restricted in purpose or lifetime, even outright disposable. The classic example is the iterator variables used in loop constructs, frequently given the single-character names i, j, k, and so on.

```
for (int i = 0; i < 10; i++) {
    printf("%d\n", i);
}
```

CODE EXAMPLE 1: C language for loop, using an iterator variable to print the integer sequence from 0 to 9.

Properly speaking, variables are a central feature of only one of the four major categories of programming languages (imperative, functional, logic, and object-oriented). The design of imperative languages—the most widely used of which include Fortran, Cobol, variants of Algol, Pascal, Ada, Perl, and Python, as well as the entire so-called C family (including C, C++, Objective-C, C#, and arguably also Java, JavaScript, PHP, Go, Rust, and Swift), but exclude Lisp, Prolog, Scheme, Haskell, and Clojure—can be understood as an abstraction of the so-called von

Neumann architecture, or more generally the stored-program computer design, which integrates a processor with memory storing both data and instructions. Understood as a name and memory address bound in a both stateful and mutable relationship, a variable represents an abstraction of a computer memory address or range of cellular addresses at which data can be stored, arranged in the sequential units called bits, nibbles, bytes, and so on.[16] The number of such units used to store data associated with a variable name will vary with memory and other aspects of hardware design as well as specifications provided by a firmware instruction set and a software operating system, among other factors; but the informal use of the term "variable" to denote a primitive data structure is licensed by the association with single units and the bounded memory segments allocated for them—for example, the one- to eight-byte numeric data types char (character), short (short integer), int (integer), long (long integer), float (floating-point number), and double (extended floating-point number) in the C language.

Though the programmer enjoys semantic freedom in choosing a variable name that is limited only by the issue of (human) readability, most languages impose orthographic restrictions that include confinement to alphanumeric characters plus a subset of other "special" characters based on those commonly used for punctuation in English; prohibiting a name from beginning with a numeric character; and prohibiting spaces within names (early versions of Fortran are a notable exception), a requirement to which the various "casing" schemes (for example "snake case," in which underscores represent spaces, and "camel case," in which word boundaries are marked by capitalization) are responses.[17] Imposed to facilitate automated parsing of program text, such restrictions have an oblique relationship to the dilated intelligibility required of placeholder words like "thingamajig," "whatchamacallit," and "whatsit," used in other language contexts where semantic precision is unavailable for whatever reason. Quite unlike the speaker of English who resorts to one of these words in a forgetfulness or ignorance momentary or otherwise, a programmer in most languages is free to choose as a variable name a nonpronounceable and utterly meaningless sequence like xxxxxxxxxx or lkjsdflkjsdflk. That the pronounceable semi-semantic words "foo" and "bar," words in the linguistic sense, are conventionally chosen instead, and that they are preferred to the direct if unspecific reference and indeed, referentiality of "whatsit," "whatchamacallit," and the like, leaves us with all sorts of interesting questions.

Four Subdomains

Ian Watson has described as "stand-ins" a wide range of linguistic phenomena in which direct but unspecific reference serves a useful or necessary role, such as the English-language use of the proper names "Jane Doe" and "John Doe," "Joe Sixpack," "Anytown," "Podunk," "XYZ Corp." and many others, including equivalents in other languages. Stand-ins, Watson argues, operate principally as "exemplifiers," generalizing an unspecified or unknown individual case or substituting a general case, such as a generalized social identity or role, for an unspecified or unknown individual. Other examples mentioned by Watson include the surname Roe used to refer to the then-unidentified plaintiff (Norma Leah McCorvey) in *Roe v. Wade*; the name Johnny, in the rhetorical question "If Johnny told you to jump off the Empire State Building, would you listen to him?"; the word "widget" as used in economic theory, as a stand-in for "all possible products"; and the range of US telephone numbers beginning with 555, left mostly unassigned and used to create fictional numbers used in film and television;[18] but one could also mention the words "thingy," "thingamabob," "thingamajig," "gizmo," "gadget," "John Hancock" (referring to anyone's signature), "Advent Corporation" (as used in legal documents), "Main Street," and "Peoria."

All stand-ins, Watson suggests, are also exemplifiers in this sense; but he identifies a special class of stand-ins (and thus a special class of exemplifiers) that perform an additional function. Calling these "metasyntactic signifiers," Watson explains that "in computer science, 'foo' and 'bar' are among the conventionalized stand-ins for 'any algorithmic variable,' and my use of the term 'metasyntactic signifier' is based on computer scientists' use of the term 'metasyntactic variable' for these stand-ins" (Watson 2005, 145–46). Metasyntactic signifiers, Watson suggests, are fundamentally algebraic in character, belonging to a set of signifiers alienated from their original or conventional referents and available for redirection to a protean class of other referents specified for the occasion. Importantly, metasyntactic signifiers serve their function only because their alienation is fully conventionalized: "if your name really were John Doe," Watson observes, "you would have to constantly emphasize to people that that really is your name and not a pseudonym" (Watson 2005, 148).

Of course, such conventionalization is a historical process, meaning that it develops over time (and can be reversed). It is subject to the geographically determined constraint of regional variation, as well. Writing in the *American Mercury* in October 1924, Louise Pound offered the following sample list of what she called "American indefinite names" catalogued in the midwestern United States: "Thingumbob, thingumabob, thingumajig, thingumajiggen, thingumadoodle, dingus, dingbat, doofunny, doojumfunny, doodad, doodaddle, doogood, dooflickus, dooflicker, doojohn, doojohnny, dooflinkus, doohickey, doobobbus, doobobble, doohinkey, doobiddy, doohackey, gadget, whatyoumaycallit, fumadiddle, thinkumthankum, dinktum, jigger, fakus, kadigin, thumadoodle, optriculum, ringumajig, ringumajing, ringumajiggen, boopendaddy, thumadoodle, dibbie" (Pound 1924, 236–37). The word "kadigin," to take one example from this list, was given a one-word definition ("thingamajig") in the first and second editions of Harold Wentworth and Stuart Berg Flexner's *Dictionary of American Slang*, published in 1960 and 1975 respectively; but it was dropped from its successor, the *New Dictionary of American Slang* edited by Robert L. Chapman.[19]

By contrast with the midwestern usage context of Pound's list, "many or most" members of which, she noted, "may be general over the United States," the context in which the metasyntactic variables "foo," "bar," and so on are used is highly constrained: a professional (and to a smaller extent, a parallel amateur) technical culture access to which requires high levels of technical ability and/or education as well as a level of privilege commensurate with it. At the same time, we might want to say that "foo," having long since reached the end of its life as a stand-in (in Watson's sense) or "indefinite name" (in Pound's sense) in conversational or written English, owes its continued life or afterlife to that restricted context, in which it is classified as a "metasyntactic variable."

The terms "metalinguistic formula" and "metalinguistic variable" can be found in technical literature as early as 1959, in John W. Backus's early formulation of what is now called the Backus-Naur form (BNF) of notation for programming language syntax, initially created for the Algol 60 specification. The following year, Martin Davis and Hilary Putnam used the term "syntactic variable" in a paper titled "A Computing Procedure for Quantification Theory," while phrasing incorporating the term "metasyntactic" appeared in other reports and documents associated with Algol 60.[20] By the 1970s, when the first versions of *The Jargon File* appeared on systems at Stanford and MIT, "metasyntactic variable" was widely used in descriptions of syntax notation and related formalisms, and no longer automatically associated with Algol alone.[21] From that point forward, the metasyntactic variable sequence *foo, bar, baz, qux* . . . developed into a characteristic feature of at least four distinct subdomains of the practice of programming.

Subdomain 1: Pseudocode

The first of these four subdomains is the writing and reading of pseudocode. Defined descriptively, as an artifact, pseudocode is a generic term for English-language (or other natural-language) writing whose syntax resembles the syntax of a particular programming language or family of programming languages, but is not identical to it. Importantly, pseudocode is *not* (yet) executable: that is, it is written for human reading only. In this sense, pseudocode is an abstraction of code in a particular programming language or family of languages, and even of program code generally. One of its main uses is to specify an algorithm, in any pedagogical, referential, or theoretical context where the core form and idea of a procedure is usefully distinguished from its implementations.

```
for j = 2 to A.length
    key = A[j]
    // Insert A[j] into the sorted sequence A[1 . . j - 1].
    i = j - 1
    while i > 0 and A[i] > key
        A[i + 1] = A[i]
        i = i - 1
    A[i + 1] = key
```

CODE EXAMPLE 2: Insertion sort algorithm described in pseudocode. Source: Cormen et al. (2009, 18).

Pseudocode can also be understood as the product of an early, preliminary stage in the process of writing a program. This is trivially implicit in the use of pseudocode to specify algorithms, insofar as algorithms have no value apart from their implementation—so that the abstraction of their specification in a textbook is best grasped as practically motivated, meant to encourage promiscuous implementation. In practice, program comments (the topic of chapter 2) may begin their life as pseudocode, as a programmer sketches out a program, then comments out lines of pseudocode while replacing them with executable syntax. But pseudocode is otherwise quite distinct from fully developed and permanent program comments, whose purpose is not to anticipate executable code but to explain it, by describing the intent of the programmer who wrote the code to another programmer reading the program.[22] (Note the comment "Insert $A[j]$ into the sorted sequence $A[1 .. j - 1]$." inserted into the pseudocode in Code Example 2, which makes this distinction clear.) We might therefore define pseudocode as human-readable text that provides a facsimile of the machine readability of program code *as well as* the nonparsable human readability otherwise reserved for program comments, whose purpose is exclusively human communication.

As an extended form of metasyntax, in this sense, pseudocode's abstraction of program code operates outside the global scope of any working program, while retaining unspecific reference to *some* program or programming language or to programmability in general. By contrast, the token-based or tokenized abstraction of metasyntactic variables like "foo" and "bar" produces neither compile-time nor runtime errors. The use of "foo," "bar," and so on in pseudocode itself, rather than in a program, might then be understood as a trace of this partial extrusion of parsing and interpretation (or parsability and interpretability).

Subdomain 2: Documentation

The second subdomain is the writing and reading of program documentation. By the mid-1970s, Jon Sachs had distinguished three forms of documentation used in software engineering: the external documentation written for users of a software application, the operational documentation written for technical service providers serving users, and the internal documentation used by programmers themselves.[23] External and operational documentation may or may not be produced by programmers (they may instead be produced by a professional technical writer whose primary task is to write documentation of an application, not to create or maintain it), but internal documentation is produced exclusively by programmers, for programmers.

An example of the latter is the documentation of application programming interfaces, or APIs. Strictly speaking, an API is a modular unit of code created by one group of programmers for use by another group of programmers, either as a software library or as a network-accessed service (or both). Its core concept, the exposure of key functionality and the abstraction of all supporting and ancillary operations, is associated with the object-oriented programming paradigm, which makes a fundamental distinction between public and private data (or state) and encourages or enforces the various forms of modularization indexed by the umbrella terms "encapsulation" and "information hiding." Today the term "API" has been generalized by the start-up culture of Silicon Valley, where it might be used, purposefully loosely, to refer to almost any software product or service at all, or, somewhat confusingly, the documentation of any software product or service, especially when documentation is generated by another software tool designed for that purpose.

Regardless of whether it carries the former connotation or the latter, however, API documentation is a form of internal documentation, intended for use by other programmers rather than for the end user of a software product or service, for whom a manual or user's guide would be prepared instead. It would be as unusual to find "foo" and "bar" used in examples in such a manual or user's guide, or even in some of what Sachs called operational documentation (depending on its level of formality), as it would be to *not* find "foo" and "bar" used in internal documentation. Such usage can be found, for example, in the API documentation of Instructure's Canvas learning management system, widely used in higher education in the United States (see Code Example 3); of MediaWiki, the PHP application that supports most of the best-known wiki websites, including Wikipedia, Wiktionary, Wikimedia Commons, and WikiLeaks (see Code Example 4); of both Plone and Drupal, two other widely used content management systems; of Amazon Web Services, Amazon.com's cloud computing platform and service; and of GitHub, the most widely used web-based hosting service for code and document repositories.[24] (Code Example 3 is clearly oriented toward unit testing, which we might consider a separate subdomain in which metasyntactic variables are commonly used, or alternately treat as another form of documentation, used by other programmers only.)

```
{ "name": "test name",
   "file_ids": [1, 2],
   "sub": { "name": "foo", "message": "bar" },
   "flag": true }
```

CODE EXAMPLE 3: Use of "foo" and "bar" in Instructure Canvas API documentation. Source: "Welcome to the Canvas LMS API Documentation" (n.d.).

If you make a request for api.php?. . . . titles=Foo|Bar|Hello, and cache the result, then a request for api.php?. . . . titles=Hello|Bar|Hello|Foo will not go through the cache—even though MediaWiki returns the same data!

CODE EXAMPLE 4: Use of "foo" and "bar" in MediaWiki API documentation. Source: "MediaWiki API:Main Page" (n.d.).

Subdomain 3. Computer Science and Programming Pedagogy

The metasyntactic variable sequence *foo, bar, baz, qux* . . . can also be found in documents and other material created for instruction in computer science and programming or software engineering. Though such usage is closely related to the reading and writing of pseudocode, as discussed above, a distinction can be made between pseudocode used for transmitting knowledge from expert to expert, for example in a reference work devoted to algorithms,[25] and pseudocode used in the instruction of novice programmers. It might be argued that the latter usage should be discouraged, insofar as a student's or other novice programmer's encounter with conventional metasyntactic variables lacks the conventionalized abstraction of their reference necessary for Watson's "stand-ins" to function as they do, which a community of experts can take for granted.

"Rant incoming, walk the other way," began an anonymous participant in Reddit's "learnprogramming" forum, devoted to discussion of programming instruction and the experience of novices, in 2014. That participant then continued:

OK, so, every time I look up an example or ask a question for some horrible reason the answer involves "Foo" and "Bar."
    When I first started out learning programming seeing those two words added an enormous amount of confusion and their repeated

use was distracting. I didn't understand their significance and was trying to relate them to the coding problem, but that was only at first.

Afterward, it made my mind run in circles because they don't mean ANYTHING AT ALL. ("'Foo', 'Bar'—Does This Actually Help Anyone—Ever?" 2014)

As the final sentence above makes clear, the author of the post "'Foo', 'Bar'—Does This Actually Help Anyone—Ever?" has understood quite well the intended purpose of "foo" and "bar" as metasyntactic variables: that is, to serve as "stand-ins" in Watson's sense. It is just as clear that the author rejects "foo" and "bar" as insufficiently conventionalized stand-ins—which have thus not yet become stand-ins in Watson's sense at all. Many responses to the original Reddit poster's reflection effectively re-described this discrepancy, either by affirming it or rejecting it. A typical example of the latter was submitted by the Reddit user "Stormflux":

Ok, but Foo and Bar are pretty standard placeholders for anyone who's been programming professionally for any amount of time. They're part of the culture. I don't see why we should change them. Do they really add that much confusion for new people?

A response to this response (which may have been submitted by the original poster) reads:

Honestly, I think they do mostly because there is nothing familiar about those words. There is nothing in foo or bar that makes one think "Yes, that's a placeholder" or "I see how that relates to a dummy example." When you're just starting out and you see single words like int, double, short, float, or def that have a special meaning, well then the first time you see foo/bar you're going to think they must have one too.

As the latter participant suggests, a novice programmer might well take such metasyntactic variables—which are similarly compressed and thus similarly epigrammatic, even gnomic—for elements of programming language syntax such as the data type specifiers int (integer) and double (a type of floating-point number) or the function declaration keyword def.

Though such complaints from learners are numerous and frequent, they are almost invariably resisted by more experienced programmers, who can often be found countering with the reasonable argument that "foo" and "bar" make code easier to read than the single-letter tokens "a," "b," "x," "y" used in mathematics (a usage also common in academic computer science instruction), or the less persuasive argument that "foo" and "bar" are usefully less culturally specific than referential nouns or proper names, to which one might otherwise be tempted to resort.[26]

Subdomain 4. Software Craftspersonship

Finally, the use of the metasyntactic variable sequence *foo, bar, baz, qux . . .* , along with a discussion of its appropriate usage and of the appropriateness of its usage at all, can be found in the numerous popular manuals of software craftspersonship that both preserve and refine the lore and tradecraft of professional software development, from Brian W. Kernighan and P. J. Plauger's *The Elements of Programming Style*, which dates to the 1970s, to Steve McConnell's *Code Complete: A Practical Handbook of Software Construction* and more recent works like Pete Goodliffe's *Code Craft: The Practice of Writing Excellent Code.*[27] Surafel Lemma Abebe and colleagues have focused the normative drives of such discourse, which can range from issues of code style to the psychology of project management and of programming itself, on the specific issue of names and naming. Among the "lexicon bad smells" (an extension of the common term "code smell," used to identify programming mistakes) they identify are "meaningless metasyntactic variables [. . .] used in an identifier. Example: foo, bar" (Abebe et al. 2011, 127). I will return to Abebe and colleagues' arguments shortly.

Meta-abstraction and the Programming Language Hierarchy

At a more general level than we have considered thus far, we might say that in all of these domains of programming the token-based linguistic abstraction of the metasyntactic variable sequence *foo, bar, baz, qux . . .* is also meta-abstractive, abstracting the abstraction of higher-level programming languages themselves, or raising it to another level. What does this mean?

Recall that the phrase "higher-level language," now widely used to describe the notation systems that emerged in the mid-1950s combining algebraic expressions with English-language keywords, refers to such systems' design for software abstraction of hardware-dependent numeric and alphanumeric operation codes. That is to say that the hierarchically higher level that such models establish supersedes a preceding abstraction, the hardware-dependent assembly language operation codes that represented binary digital instructions using alphanumeric mnemonics. Purely numeric so-called machine code, representing hardware processes for storing and incrementing or decrementing simple register values (achieved using electrical polarization or other means, but not so far removed from counting with sticks or other tokens), is thus abstracted by the alphanumeric mnemonic abbreviations of an assembly code created by a processor's manufacturer for convenience in writing machine instructions. Assembly code is in turn abstracted by higher-level languages that provide a programmer with a lexicon of everyday English-language words mapped to ranges or aggregates of instructions in machine code.

If it is to be meaningful to speak of a supersession of higher-level languages themselves, in such a hierarchical schema, we might do well at this point to reject the science fiction of the usual suspects, from fifth-generation projects to generalized artificial intelligence, and admit the enclosure of computation within natural language that programming languages represent, and that metasyntactic variables might be said to index.[28] In a way, we might say that in context, the meta-abstraction of the sequence *foo, bar, baz, qux* . . . makes the abstraction of the programming language hierarchy tangible, lending it a concreteness it cannot possess, but which is useful as a mode or artifact of (self-)observation. "Foo" functions as a nonsemantic keyword within a domain of English-language reserved words whose semantics are not those of the English language, but of a programming language design. That is to say that it is a token name, belonging to the category of names the choice of which is granted the programmer, such choice being limited not by the minuscule bounds of a programming language's syntax (its reserved words), but by the notably vast bounds of the English language itself. And at the same time, it is the token name that the programmer chooses in that subcontext in which English-language meaning is not only not required, but is unwanted.

"Experienced programmers," Abebe and colleagues observe, "choose identifier names carefully, in the attempt to convey information about the role and behavior of the labeled code entity in a concise and expressive way. In fact, during program understanding the names given to code entities represent one of the major sources of information used by developers" (Abebe et al. 2011, 125). Their taxonomy of what they call *lexicon bad smells* in software includes categories of outright programmer error, such as misspelled identifiers or identifiers clearly used in the wrong context (such as a module or package other than the one to which a segment of code belongs). It also includes categories of effects produced by carelessness—inconsistent use of identifers; unusual grammatical structure; redundancy, in integrating data type indicators into names when a language requires them to be declared explicitly anyway—or by haste, such as extreme contraction (for example, using single-letter identifiers). Deeper problems are indicated by the overloading of identifiers or other forms of synonymy, which conflate multiple roles or functions that would be better separated; or by elided hyponymy and hypernymy in class hierarchies, in an object-oriented design.

About the use of metasyntactic variables, Abebe and colleagues have little to say beyond categorizing them as "meaningless terms" used hastily or carelessly. The lack of sustained attention suggests that in their view, the odor of *foo, bar, baz, qux* . . . is relatively mild. Though the meta-abstraction of the metasyntactic variable sequence is left unexplored, it is arguably key to the nature of the distinction between the activity of writing both readable *and* executable program code, on one hand, and activities that involve reading that code with human understanding, on the other—a distinction at the heart of Abebe and colleagues' study of how one of these domains impacts the other. *Foo, bar, baz, qux* . . . are "meaningless" in both the domain of natural language and the domain of programming language sytax, and their use as names chosen freely by the programmer reinscribes and reminds us of the difference between natural language and code, just as does any other freely chosen name. If their conventionalized use as meta-syntactic signifiers tends to attenuate that difference, for example in encouraging novice programmers to mistake them for reserved words, it is also true that, as we have seen, "foo" has its own specific and determined, if not unambiguous or especially well documented natural linguistic history and derivation. To echo the title of RFC 3092, we have an etymology of "foo."

Cultural Pointers

Another way to understand the cultural function of the metasyntactic variable sequence *foo, bar, baz, qux* . . . is by analogy to the technical function of the pointer. In programming language design, a pointer is an object storing a memory address. Where data is linked to variable names, a pointer variable is used to point to a memory location other than the location that would otherwise be linked to a name, when assigning data representing primitive values; when linking a name to a data structure distributed in memory, a variable may be a pointer variable by default. (While it was by no means the first to formalize the pointer as a feature, the C language is, of all the most widely used programming languages, the one most strongly identified with it, very much for better and for worse.) Many forms of data processing require that data not be changed in or by the process of evaluating it, if its results are to be relied on. While such a scenario might be imagined unusual, the truth is the opposite: in the stored-program paradigm expressed by all general-purpose computing architectures today, such operations are intrinsic and extensible even to program code itself, treated as just one more aggregation of data, and early computers routinely modified program code while running a program, without any of the many safeguards later developed to reduce its inherent risk. Pointer variables, understood as aliases, preserve something of this volatility in their abstraction of the memory addressing expressed by variable assignments: a virtualization that in setting free the reference of a token offers the useful illusion of direct memory access.

Pointers provide efficient access to memory contents by multiplying the paths that can be followed to access them. Risky, yet handy in the small-scale, improvisational systems programming for which the venerable C language, for example, was designed (that is, to implement the Unix operating system), they are also widely used in large systems written mostly or entirely in C++, C's successor. In addition to being difficult to use properly, they are easy to exploit to gain access to areas of memory storage that need protection, such as the areas used by an OS and the startup programs that load it. For this reason, pointers are features of relatively few languages in use today (in addition to C, this includes Objective-C, C#, Go, and Swift as well as C++). By contrast with the general concept of *reference*, meaning any controlled and authorized path to extant data used instead of a new object and allocation of memory, pointers live up their name, providing access to a "point" in memory—perhaps especially "pointed," even "pointy" access—whether that point is safe to access or not. A pointer is purely functional in this sense, ordinary or extraordinary depending on context, like an exterior door on an upper floor of a house.

If not in their alienation from what original natural-linguistic referents they once had, then in the extensible alienability in which they remain available for redirection, the metasyntactic variables *foo, bar, baz, qux* . . . function as *cultural* pointers: specifically introspective instances of a more general concept of reference that governs the contact (or no-contact) zone in which natural languages and programming languages appear to touch. In the incomplete but realizable conventionalization that encourages novice learners to confuse them with reserved words, they are not "meaningless," as Abebe and colleagues would have it, but meta-abstractive—not unlike Watson's "self-destructive" stand-ins, whose conventionalization is a second-order, rather than a primary process. "Designers who create stand-ins and pseudonyms meant to be convincing," Watson observes,

> can, if they wish, build a 'self-destruct' feature into this convincingness, so that it becomes obvious after a little reflection that the stand-in does not actually exist. For example, any American road atlas will confirm that there is no interstate highway 13 in the United States (perhaps for the same reasons that many tall buildings in America have no 13th floor). (Watson 2005)

Where the absence of a road or a building floor numbered 13 represents the modification of a planned technical-material system to accommodate something potentially resistant or adversative (here, the ethnographic concept of superstition) in its culture of use, the residual lexicality of the metasyntactic variable is a cultural space in the planned technical system of a code. Retaining both semantic history and semantic availability, as the always already dereferenced sequences xxxxxxxxxx or lkjsdflkjsdflk cannot do, though they may with as much permission be chosen as names, *foo, bar, baz, qux* . . . point simultaneously or bicamerally to human language, as a technical system *and* a culture, and to the ordinally mapped virtual space of stored data on which pointer variables operate in redirection. And whereas when dealing with code, the term "metasyntactic" serves to separate such tokens from the specified, thus valid syntax of a programming language, when dealing with language, they are better understood as re-representing a linguistically derived code's abstraction of language. Watson's two examples do not actually illustrate the same thing: any building more than twelve stories high *has* a numeric thirteenth floor, regardless of whether that floor is labeled as such, whereas we certainly can say that no US Interstate Highway System 13 exists (although one was proposed by North Carolina state senator Robert Lee Humber in 1964). It is perhaps less useful to schematize the difference, here, than to take it as demonstrating the volatility of exemplarity itself, in this context.

Consider a similar artifact of the culture of programming, found just as often in software documentation and instructional materials, if not in pseudocode. The phrase "Hello world" refers to a demonstration program that does nothing beyond outputting the English-language phrase "Hello world" (or its common variations: "Hello, world," "Hello world!," and so on). A "hello world" program is what we might call purposively purposeless, in that its only use is to serve as a minimal demonstration of programming language syntax or a minimally executable program. As such, it shares with the metasyntactic variable that "meaninglessness," in the sense of Abebe and colleagues, that we have suggested is better understood as a higher-order or meta-abstraction. Call it "meta-procedural," in this case, since unlike the metasyntactic variable, which performs its technical function by remaining asyntactic, a "Hello world" program has no value if it cannot be successfully compiled or directly interpreted: that is, if it fails to output the string "Hello world."

"Hello world" is also very often the first program written by a novice. This gives it a second characteristic that distinguishes its function quite decisively from that of the metasyntactic variable sequence and its presumption of expertise, even neatly inverts it. For the string "Hello world" functions as a greeting—a greeting produced by an agent who is deliberately ambiguous or confused. The text "Hello world" is written by the computer, so to speak (more precisely but not quite as exactly, by the compiler or interpreter of the program), as a greeting to the world, as if the computer had just been born, or had just acquired the ability to write. But in its pedagogical function as a novice programmer's first program, "Hello world" also speaks for that novice programmer, acquiring and exercising the newfound power to (apparently) directly control a machine. In linguistic terms, the English-language phrase "Hello world" is phatic or performative, establishing a social relation rather than communicating a message; whereas in technical terms the "Hello world" program produces only a so-called "side effect"—both literally, in its direction of output to an output device, and figuratively (or generalizably) in what I have called its purposive purposelessness. Grasped in a still broader context, the context established by sociolinguistic theory, "Hello world" is dialogic, meaning not so much the turn-taking of two communicators focused on transmitting and receiving a message, as the polyphony of two speakers speaking at once, and saying the same thing, to an interlocutor who is in each case one's self, the other, and an implied third party ("the world") at the same time.

```
#include <stdio.h>

int main(void) {
    printf("Hello world!\n");
    return 0;
}
```

CODE EXAMPLE 5: "Hello world" program in C.

If "Hello world" thus enacts a kind of primal scene of computing, in the novice programmer's embrace of power over an electro-mechanical servant, even a prosthetic slave, the metasyntactic variable "foo" stands in some way for that power's conventionalization and for the programmer's habituation to it. "Foo," we might say, marks the authority, even the authoritarianism of computing in a manner homologous with "snafu," the widely used and fully conventionalized token of specifically US military slang from which some have imagined the former derived. The function of "snafu," Frederick Elkin has written, is to *compliantly* caricature military authority: it is used by the US soldier in a resigned acceptance of both his requirement to obey *and* the disarray of the authority that requires his obedience. This "caricaturing of both the Army and himself," Elkin concludes, "evidences an adjustment in which the soldier accepts his subordinate position in his own mind but does not completely adopt the subordinate role" (Elkin 1946, 422).

This may seem to take us far afield from RFC 3092 and its proposed etymology. But if we accept, as I am suggesting that we accept, that the etymology of "foo" leads us not only deep into the history of programming but back out again, so to speak, into its broadest possible, indeed meta-abstractive context, then we may have to conclude that much about the power to "speak code," as that power is promoted and pursued today, still demands closer examination.

———

Along with the previous chapter, this chapter has focused on a contact zone within a computer program where natural languages and programming languages appear to touch or intermingle. In the next two chapters, I turn to the close study of two specific programming languages. One of these, Snobol, is now almost entirely obsolete, while the other, JavaScript, has become the most widely used programming language in the world. What they have in common is their technological-historical roles as mediators of the relationship of professional technical expertise to casual, inexpert, and non-technical uses of computing. This role developed accidentally for Snobol, and was programmatic in JavaScript's original design, yet in each case, that role became an inversion of an original imagined and projected purpose. This makes Snobol and Java-Script useful candidates for demonstrating how the philological study of programming languages and their usage cultures might proceed. While we will resume our discussion of automation more directly in the final chapter of this book, in the meantime, Snobol and JavaScript together afford us a view of a byproduct of the technical logic of software as the automation of automation: namely, the opportunity for novice programmers to experiment with automation for the first time.

FOUR

Snobol

A REMEMORY OF PROGRAMMING LANGUAGE HISTORY

Orthogonality

"The SNOBOL languages," Ralph E. Griswold wrote in a memoir of his work as their designer and most prominent public representative until his death in 2006, "have had little apparent direct effect on the mainstream of programming language development" (Ralph E. Griswold 1981, 634). What Griswold called the "orthogonality" of Snobol has been understood in different ways by Griswold himself and by other commentators on this interesting but submerged episode in the history of programming languages, two case studies in which are presented in this chapter and the next. A so-called right-brain language (that is, one supposedly designed for inductive and creative applications) in an era when software development was understood to need radical procedural and managerial discipline above all else, Snobol was presented as relatively unconstrained and deliberately easy to use at a moment when the so-called software crisis was pushing programmers and their managers in the opposite direction. Snobol's untimeliness, in this sense, has combined with the orthogonality of its design to buttress its reputation as a particular kind of novice's programming language. That is to say that when it is remembered at all, Snobol is often remembered as a language for experts whose specific expertise is not in computing, and is not technical in the usual sense: that is, which lies in the nontechnical knowledge disciplines, especially the humanities.

In this chapter, my argument is that the orthogonality of Snobol rewards reexamination and reflection from a perspective that integrates the history of programming languages into a broader cultural history of technical expertise in computing. The orthogonality of Snobol cannot be satisfactorily explained, I suggest, by appeal to its imagination as a humanities programming language; indeed, Snobol appears to have acquired both that domain of use and the reputation that went with it quite accidentally. Rather than a direct recovery of past practices in nontechnical or humanities computing as such, then, a reexamination of Snobol's history offers a specific, angular vantage from which to consider the relationships among technical design, use cases and usage domains, domains of knowledge and modes of expertise.

Background

No single document in the general history of computing is as closely associated with the emergence of so-called structured programming as Edsger W. Dijkstra's polemic "Go-to Statement Considered Harmful," originally published in 1968 (Dijkstra 1968). Perhaps the most common form of written computer program in the late 1950s and early 1960s was a purely linear sequence of instructions, intended to be interpreted in their order of arrangement—that is, descending from the first instruction to the last, by analogy to conventional reading. Branching, or choosing between multiple possible outcomes based on some condition, and iteration, or repeating the same action, were performed by jumping forward or back across that linear sequence using so-called go to statements, which instructed a code interpreter to jump directly to a specifically labeled line of the program that was *not* the next line in the existing sequence—and perhaps later to jump back. Emerging from struggle with the difficulty of reading, understanding, and modifying such programs after they were written, structured programming emphasized use of the compound statements or block structures introduced in Algol 58 and subsequently incorporated into other languages. While as an attempted solution to the problem of so-called spaghetti code (a phrase used to mark the illegibility of a crudely nonlinear execution path through a linear sequence of instructions), structured programming stood for a new approach to programming with articulated principles and a fundamental theorem (Böhm and Jacopini 1966), it was often represented by the methodological shorthand of a prohibition on go-to statements.

```
10 PRINT "HELLO"
20 PRINT "GOODBYE"
30 GOTO 10
```

CODE EXAMPLE 1: Go-to statement in an "unstructured" program in the Basic language, forming an infinite loop that will display the words "HELLO" and "GOODBYE" in succession until it is interrupted.

```
while (1) {
    printf("HELLO\n");
    printf("GOODBYE\n");
}
```

CODE EXAMPLE 2: The "structured" equivalent of Code Example 1 in the C language, written as a block enclosed by the brace characters { and }.

As a feature, the go-to statement (hereafter "goto") was available in most widely used programming languages at the time, including many that made no special claims of approachability or ease of use by novices or nonprofessionals. Dijkstra's denunciation of gotos is among the more widely circulated of the many and varied occasional writings in which he advertised his wit and contrarian disposition, and of which he left a vast ancillary archive alongside his technical contributions to computer science. As a casual, yet fundamental document of the cultural history of computing and computer programming, it is often associated with Dijkstra's equally conspicuous fulminations against early programming languages like Cobol and Basic and their use in classroom programming instruction in particular. That Dartmouth BASIC succeeded in its main purpose, to provide all Dartmouth students with experience writing simple programs, and would go on to play an important role in the emergence of the personal computer and an early tradition of expressive computing,[1] is something one can affirm against Dijkstra's antipathies while accepting his impatience with the possibly unnecessarily primitive affordances of the time.

When Griswold suggested in 1981 that Snobol's "lack of direct influence" on the history of programming languages "is probably a necessary consequence of the direction SNOBOL has taken, which is deliberately contrary to the mainstream" (Ralph E. Griswold 1981, 634), the "mainstream" was represented by the C language, developed along with the Unix operating system and today one of the most widely used programming languages in history, still widespread after half a century of continuous use. Designed with the principles of structured programming in mind, and implemented by working programmers rather than by academics or committees, C "is considered by many to have marked the beginning of the modern age of computer languages," presenting solutions to many of the problems that preceded it (Schildt 2017, 4). While Dijkstra's indictments of the "infantile disorder," "fatal disease," "mental mutilation," "criminal offense," and "mistake carried through to perfection" of Fortran, PL/I, Basic, Cobol, and APL respectively (Dijkstra 1982, 130) may be comical in their vehemence, the frustration they express helps to explain the rapid spread of C from the late 1970s onward.

In this context, while Snobol's orthogonality as Griswold imagined it can be understood as of a piece with *any* of Dijkstra's targets in "How Do We Tell Truths That Might Hurt?," which were soon to be collectively eclipsed by C, it might be grouped specifically with Basic, a novice's language whose design for students in technical disciplines (Kurtz 1981, 518) ensured indifference to the needs (and vexations) of technical professionals and scholars in the technical sciences alike. Since the nontechnical disciplines include the humanities, a knowledge domain often imagined to be incommensurable with the technical sciences, one might suggest that the refusal of structure represented by goto statements was a feature, not a bug, enhancing the accessibility of Basic and Snobol to programmers in nonscience disciplines. This would not be entirely misguided. Like Basic, Snobol expediently and unapologetically embraced the atavisms and convenience features that Dijkstra indicted as "disastrous" and which he demanded be "abolished from all 'higher level' programming languages" (Dijkstra 1968, 147).

However, two problems with this coupling emerge, to give context to which is only part of my goal in what follows. The first is that unlike Basic, conceived for the curriculum of a college devoted to undergraduate student education, Snobol was, like the C language, developed as an internal project at Bell Laboratories (today Nokia Bell Labs), the private telecommunications research center that played a central role in the his-

tory of computing. This meaningfully different provenance complicates
the question of essential affinities considerably. The second difficulty is
that when one follows its history, Snobol's reputation as a humanities-
oriented programming language appears to have been acquired almost
entirely by accident, and certainly not by design. This means that the
orthogonality of Snobol is not—or is not simply—the orthogonality of the
relationship of humanities research questions and practices to answers
and applications devised by computation (a discrepancy that need not be
a matter of dispute). Its story includes some interesting twists and turns.

Strings

"SNOBOL is responsible," Griswold argued, "for introducing to high-
level languages the idea of the string of characters as a single data object
rather than as a workspace or an array of characters. This is a concep-
tual matter, and one of considerable importance, since the treatment of
a varying length string as a single data object permits a global view of
string processing that character aggregates inhibit" (Ralph E. Griswold
1981, 626). This was no exaggeration. No type of human-readable data,
not even the apparently most primitive representation of numeric data,
is easy to represent using a computer, and the representation of natural-
language text is difficult in proportion to its importance (between numeric and textual applications of computing, the latter outnumber the
former, and the results of the latter need representation in some textual
form to be useful—there are no purely numerical end-to-end applications
of computing).[2]

In programming language design, the word "string" refers to a form
of text data consisting of a sequence of representations of written characters in a natural language. A string may contain anywhere from zero
such characters to however many constitute the upper bound on data
transmission and storage imposed by the hardware architecture of
a given system. While a string may well happen to represent what in
natural-language terms we call a letter, a word, a phrase, a sentence, a
paragraph, and so on, in such cases the coincidence is meaningless, as
the concept of a string is quite different.

Crucially, a string may include so-called white space characters separating words or other units of text in many languages. Thus not only
"T" and "The" but also "The quick brown fox" and "the quick brown fox

jumps over the lazy dog" are strings. Though under most circumstances, white space characters are no more visible in a text displayed on an electronic screen than they are on a page of printed text, they are representations nonetheless, rather than merely dividers or segmenting devices for organizing a representation of writing. In the Unicode character encoding system used almost universally today, the usually nonvisible symbol assigned the identifier U+0020 SPACE is assigned to the space bar on keyboards, which when pressed does in fact cause a character to be displayed, just like any key visibly labeled with a written symbol. This is quite different from the mechanical or electromechanical spacing action of a typewriter, in which the space key, rather than creating some representation on paper, simply moves the writing head instead, much as the handwriter holding a pen moves to leave space between the end of one word and the start of the next.

The specific challenge in storing and representing text data lies in the need to preserve information about the proper (that is, legible) sequence of written characters. From the earliest version of the Fortran programming language onward, the solution was to use a data structure called a linear array, a group of blocks of memory with sequential addresses. In a so-called character array, a linear sequence of blocks of memory allocated for representations of the written characters in a unit of natural-language text, the sequence of memory addresses represents the sequence of characters in the string, and thus the sequence of letters (or other written symbols) in printed text or text on a display screen. The disadvantage of using such a "bare" array to store text data as a sequence of individual characters is exposed by the importance of string length in many applications of computational text processing. As an allocated unit of memory, a basic linear array is a static data structure: that is, only as much memory as is needed to represent the text "The quick brown fox" is allocated, when that array is created, which means that if one wants to perform certain kinds of operations on that string, like removing or adding one or more characters to produce "The quick brown" or "The quick brown fox jumps," the result of those operations will no longer fit neatly within the memory space originally allocated for the string. Adjustments must be made, either to prevent leftover character data from polluting the result of such changes, or to find the additional memory needed to extend it (in practice, this means allocating more memory and copying the result to it, then discarding the original).

0	1	2	3	4	5	6	7	8	9	10
M	e	r	k	k	i	j	o	n	o	\0

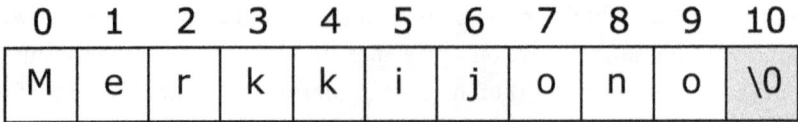

FIGURE 5. Generic representation of a character array. Source: Wikimedia Commons.

For it to be practicable, meaning that programs to perform it can be written with relative ease, text processing requires flexibility with respect to the length of strings, and this is a problem that Snobol's designers set out to solve from the start. Their solution is now something close to universal. It involved bundling information about the string, such as its length, with the data used to represent it, and in later developments, with special functions used to perform commonly desired transformations, such as transforming a character (for example, from lowercase representation to uppercase representation), replacing one character with another, or extracting a specified portion of a string. This form, in which the data representing text is bundled with information about it and functions for operating on it, is most commonly called an object. Again, Griswold was not exaggerating when he suggested that "the freedom that string data objects provide the programmer is enormous" (Ralph E. Griswold 1981, 626): the representation of strings as complex objects is a canonical instance of the abstraction, or layering of programmers' conveniences, that gives meaning to the phrase "higher-level programming language."

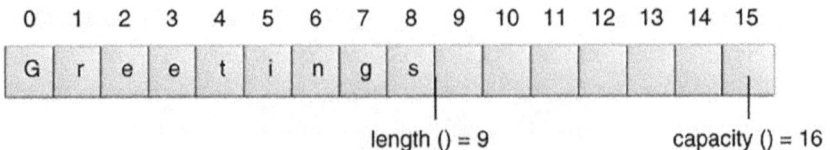

0	1	2	3	4	5	6	7	8	9	10	11	12	13	14	15
G	r	e	e	t	i	n	g	s							

length () = 9   capacity () = 16

FIGURE 6. Representation of a Java language StringBuilder object including measures of length and capacity provided by object methods included in the class. Source: The Java Tutorials: The StringBuilder Class. Copyright 1995, 2022 Oracle and its affiliates. Used with permission.

Precursors: COMIT and SCL

Though Griswold himself has a complicated view of Snobol's gene-alogy (see below), it is commonly understood to have been derived at least in part from the COMIT language, and Griswold has acknowledged COMIT's influence. COMIT was developed during the mid-1950s by the Mechanical Translation Group of the Research Laboratory of Electronics and the Computation Center at MIT, in a project directed by Victor H. Yngve. Its origin is in early work on machine translation (henceforth "MT"), specifically the theoretical and "perfectionist" approach of that work as it was pursued at MIT and contrasted with the empirical and operational approach of other research centers during the 1950s (Lennon 2014, 139). As Yngve put it in 1957:

> The current M.I.T. approach to mechanical translation is aimed at providing routines intrinsically capable of producing correct and accurate translation. We are attempting to go beyond simple word-for-word translation; beyond translation using empirical, ad hoc, or pragmatic syntactic routines. The concept of full syntactic translation has emerged: translation based on a thorough understanding of linguistic structures, their equivalences, and meanings. (V. H. Yngve 1957, 59)

The next year, in an article in the journal *Mechanical Translation* titled "A Programming Language for Mechanical Translation," Yngve described a "notational system for use in writing translation routines and related programs . . . specially designed to be convenient for the linguist so that he can do his own programming" (Victor H. Yngve 1958, 25). Two issues are especially salient here. The first is Yngve's emphasis on empowering the researcher, specifically the professional linguist, to program the computer without assistance. Linguists, Yngve wrote, were forced to rely on programmers to produce custom programs for working with their data; to the basic inconvenience of this division of labor was added that of the incommensurabilities of equal but distinct modes of expertise. "The linguist has not become aware of the full power of the machine, and the programmer, not being a linguist, has not been able to use his special knowledge of the machine with full effectiveness on linguistic problems" (Victor H. Yngve 1958, 25).

But the occasion for the newly empowered linguist-programmer's encounter with automated computation is even more fundamental, in the adaptation of programs for computers originally "designed to handle mathematical problems" to the processing of natural-language text data. The "notational system" presented by Yngve to meet both of these needs—processing text data, and doing so using a system the linguist could master—was called COMIT, and the introductory textbook published three years later by the Research Laboratory of Electronics and Computation Center directly echoed Yngve's two principal concerns:

> Although the automatic digital computer has been designed specifically for numerical procedures, there is an ever widening group of non-numerical procedures to which it is being applied [. . .] In the past, the preparation of programs was often relegated to a programming specialist who acted as an intermediary between the user and the machine by translating the user's problem description into an appropriate detailed program [. . .]The purpose of the COMIT system is to minimize or eliminate the delays and the barrier by providing the user with a programming language that is so simple to use and so well suited to his task that he can prepare the programs himself. (*An Introduction to COMIT Programming* 1961, 1)

A primer for nonprogrammers, *An Introduction to COMIT Programming* was followed by the corrected second edition of *COMIT Programmers' Reference Manual,* which described COMIT as "a problem-oriented automatic programming system that is designed to be convenient for writing programs involving non-numerical processing of information," available for the IBM 709 and 7090 computers, with plans to make it available for the IBM 704 (*COMIT Programmers' Reference Manual* 1962, 1). That year, in an article in *Communications of the ACM,* Yngve described COMIT as a general-purpose programming language that had outgrown its specialized origins in MT research. It managed memory for the programmer, rather than requiring the programmer to do it manually, and its syntax, Yngve argued, had been designed to be "easy and natural to use, and easy to learn" for the "individual research worker." Along with MT, its potential applications included "information retrieval research, vocabulary analysis, text processing-editing, random generation of English sentences, automatic milling machine programming, sociological data reduction, simulation of human problem solving, simulation of games, theorem-proving and mathematical logic, logico-semantic investigations, [and] electrical network analysis" (Victor H. Yngve 1962, 19).

Derived from the so-called rewrite rules (phrase structure rules) used in Chomskyan generativist linguistics, COMIT's syntax, it was hoped, would also be accessible to researchers in fields outside this rather technical subdomain of linguistics. Indeed, by 1964 Yngve was boasting that "COMIT can be taught using [*An Introduction to COMIT Programming*] in six one-hour lectures [. . .] Students that graduate from this course have programmed a rather sophisticated problem, translating from English into Pig Latin" (Victor H. Yngve 1963, 84). In her monumental early history of programming languages, Jean Sammet was more restrained, remarking that the COMIT user was "primarily intended to be a nonprogrammer and, in particular, a linguist": COMIT, she observed, "was designed to provide the professional linguist with a programming system in which he could easily write the programs that he needed for his research," though she acknowledged that "it has actually received usage outside this area" as well (Sammet 1969, 419, 417). Such usage was not, however, more technically specialized than linguistics; if anything, wider COMIT adoption had moved in the direction Yngve suggested. This was consistent with COMIT's place in an emerging constellation of incipient general-purpose programming languages: "The language definition itself," Sammet noted pointedly, "is written in English, with no attempt at rigorous formalized notation, and there is considerable ambiguity in some places" (Sammet 1969, 420)—a retrospective observation reminding us that COMIT was first described by Yngve in 1958, the same year in which appeared the first of the Algol language specifications, later famous as largely unimplemented formalisms and repurposed notation systems (chiefly the so-called Backus-Naur form, which was deliberately divorced from Chomsky normal form).[3]

Describing COMIT as "the first of the string processing languages," Sammet suggested that Lisp, variants of Algol, and other languages had adopted its model for string pattern matching and transformation procedures, which "did not exist anywhere else" (Sammet 1969, 435). Ironically, perhaps, given Sammet's remarks on its specification, COMIT program syntax was regarded as unintuitive, something Sammet felt obliged to acknowledge. "COMIT is a language," she wrote, "whose format is unlike most others [. . .] At first glance [it] appears highly formal and difficult to read and write. In actual practice, it is fairly simple in notation, compact, and both succinct and quite natural within its class of applications" (Sammet 1969, 417–18). In an attempt to resolve the contradiction, Sammet extended this argument into a contrast with Lisp, a theoretically rather than practically motivated academic language associated with artificial intelligence research at MIT and Stanford. "One of the most astounding things about COMIT," she reported, "is the discrepancy between its apparent and surface difficult notation and the actual ease of writing and using the language. This contrasts sharply [. . .] to LISP, whose notation is inherently simpler than COMIT but which seems to be much harder to learn and to use" (Sammet 1969, 435).

```
FIND BOY + $ + , + $ = 2 + 3 + GIRL // *WSMI 2, NEXT
```

CODE EXAMPLE 3: A COMIT "rule," or complete expression, performing the following action: determine if a string contains the word "boy" followed by any number of words, then a comma, then any number of words again; if found, remove the word "boy," insert the word "girl" after the comma, and discard the rest of the string; then write out the text up to and including the comma, and move on to the next rule. As printed in Sammet (1969), 417.

Rather than as commands in the conventional sense (that is, imperative or direct specifications of steps in a procedure), COMIT instructions were expressed as descriptive "rules" for the transformation of string data. Conventional symbols for arithmetic operations, particularly the addition symbol or plus sign (+), were repurposed for string-processing operations such as joining two strings into one; meanwhile, arithmetic manipulation of numerically represented data, an almost entirely secondary feature in COMIT, was offloaded to infrequently used symbols. Consistent with this demotion of arithmetic, COMIT provided no assignment syntax, a characteristic feature of imperative programming languages in which the assignment of a data value to a constant or variable name mimicked (while repurposing) algebraic notation. Together, at least according to Sammet, these design decisions represented COMIT's primary shortcomings.[4]

```
INTEGER A, B, SUM
A = 1
B = 2
SUM = A + B
```

CODE EXAMPLE 4: Assignment statements in Fortran IV, contemporaneous with COMIT and designed (and still used today) primarily for numeric applications.

A second influence on Snobol, according to Griswold and his collaborators, David J. Farber and Ivan P. Polonsky, was SCL (Symbolic Communication Language), developed by Chester Lee at Bell Labs in 1962. SCL was designed for processing symbolic expressions, and its basic semantic unit was a line of text, imagined as corresponding equally to an algebraic expression (or equation) and a natural-language sentence. Though it was similarly flexible to COMIT, in that respect, SCL's applications at Bell Labs were quite different: not natural-language representation and manipulation, but symbolic mathematics. As Griswold put it:

> We were working at Bell Laboratories in the early 1960s on developing programs for manipulating mathematical formulae[. . .] At that time, the available programming languages were designed primarily for numerical and business data processing; facilities for doing general-purpose string manipulation were crude. A predecessor of SNOBOL, called SCL, had been developed at Bell Labs to meet the specific needs of the research, but SCL was very slow, awkward to use, and the space it provided for string data was severely limited. It was sheer frustration that motivated us to develop a new programming language. (Ralph E. Griswold 1985, 1)

Snobol: Origins

Though Snobol's creators have emphasized that it was developed internally at Bell Labs for their own research needs first of all, they also generalized those needs as typifying evolution in the applications of computing. As they put it in 1964, "The ability to manipulate symbolic rather than numeric data is becoming increasingly important in programming [. . .] It is clear that more general symbol manipulation languages will materially expand the class of problems that can be programmed with reasonable time and effort." In selecting the string as the fundamental data structure in Snobol, they reported, they had been guided by the conviction that "most symbol manipulation problems of current interest may be naturally described in terms of string manipulations" (Farber, Griswold, and Polonsky 1964, 21).[5] Though in *Programming Languages: History and Fundamentals* (which does not mention SCL) Sammet positioned Snobol in a genealogy of string- and list-processing languages with COMIT as its immediate ancestor, Griswold insisted that SCL was "a much more immediate influencing factor," even if that influence was obscured by the fact that "no information about SCL was ever released outside BTL" (Ralph E. Griswold 1981, 615). Griswold presented Snobol's design as a synthesis of SCL's orientation to "algebraic manipulations on parenthesis-balanced strings of characters" and COMIT's origins in MT research, which made natural-language words the "constituent" units of operation. Both SCL and COMIT, that is to say, had been designed for symbol processing, being "essentially single data type languages in which numerical computation was awkward," but their domains of application were otherwise quite different (Ralph E. Griswold 1981, 616).

Snobol was designed in the Programming Research Studies Department (PRSD), a division of the Switching Engineering Laboratories at Bell Labs. Griswold notes the ambiguous position and structure of the PRSD in 1962, as a group of six members with no formal internal hierarchy, neither fully integrated into its host division nor affiliated with the separate Research Division. Its head, Chester Lee, was in Griswold's words "mainly concerned with his own research and other members of the department worked relatively independently" (Ralph E. Griswold 1981, 601–2). Applications of Lee's SCL being explored in the PRSD at the time included automata theory, graph analysis, and formula manipulation, including factoring multivariate polynomials and analyzing Markov chains, and Farber, Griswold, and Polonsky found themselves frustrated by the limitations of their supervisor's design. In a development that might be unusual in a more structured context, the three researchers abandoned the programming language designed by their group's supervisor and began experimenting with COMIT instead.

COMIT itself was soon abandoned, however, for reasons that are interesting in light of the wide range of later applications of Snobol, the language that Farber, Griswold, and Polonsky would design to replace it. As Griswold put it, "Although COMIT was interesting, its difficult syntax and orientation toward natural-language processing made it unsuitable for formula manipulation and graph analysis" (Ralph E. Griswold 1981, 602). This remark suggests that initially, at least, Griswold had imagined Snobol as a better SCL, more suitable for the work of Lee (1961) on path-connection problems, intended for application in logical drawing, wiring diagrams, and other routing tasks, and for Griswold's own work on flow graphs (R. E. Griswold 1962) and his work with Polonsky on finite Markov chains (R. E. Griswold and Polonsky 1962), and that given its design for natural-language processing, which was not an interest of the PRSD researchers at the time, COMIT represented a step in the wrong direction. In Griswold's words, "COMIT [. . .] was designed primarily for linguistic processes, and it had a concept of constituents really intended to be words, which didn't help a great deal with algebraic expressions" (Ralph E. Griswold 1981, 647). While it is tempting to imagine that Snobol emerged from a firm conviction that strings, the fundamental data type in natural-language processing, provided the best implementation of certain forms of symbolic computation, it is not clear that any of Snobol's designers actually held such views, at least in advance. Rather, it may be that the freedom enjoyed by the PRSD researchers permitted and even encouraged improvisation, and that Snobol's subsequently developed affinities and applications represent a loop rather than a straight line.

Reflecting its closer relationship to SCL, the first name given to what would later become Snobol was SCL7 (the current numeric version of SCL at the time being 5). The next, given to the first working implementation, was SEXI, for *S*tring *EX*pression *I*nterpreter. It appears that the association with the homophone "sexy" was embraced, rather sophomorically, by the language's designers. Indeed, this was the first in a series of more or less whimsical episodes, making for entertaining anecdotes, by which Snobol comes to seem like a cultural antecedent of Python, the general-purpose programming language whose name was chosen in tribute to the British sketch comedians Monty Python:

> We struggled for a long time for a name. We spent more time trying to find a name than we did designing and implementing the language. And I'm not exaggerating. And I don't think that anyone should downplay the amount of effort it takes to derive names or the unfortunate consequences that they may have [. . .] we felt that the acronyms that had been given for programming languages reflected a lack of maturity in the computer field. Everybody wanted to be cute. And we struggled for a long time to try to match that immaturity [. . .] my recollection is that I went into the office of Dave Farber and Ivan Pulonsky and I say, "Well, I've really got a disgusting name." And for some reason they like it, and I think at that time—my recollection is that Dave Farber says, "That sounds good because it doesn't have a snowball's chance in hell." (Ralph E. Griswold 1981, 657)

In the end, "Snobol" was a pseudo-acronym or "backronym" deliberately formulated in the spirit of satire, as indicated above. "As I recall," Griswold reflected, "I came up with the name first and then put together the phrase from which it was supposedly derived—StriNg Oriented symBOlic Language." Though it stuck, this choice would be the occasion of no small amount of regret, for Snobol's creators: "Had we realized the extent to which SNOBOL would become popular," Griswold remarked, "we probably would have selected a more dignified name. I recall that it was years before I could give a talk on SNOBOL without a sense of embarrassment about the name" (Ralph E. Griswold 1981, 605).

Chroniclers of the history of programming languages, beginning with Sammet, who regard Snobol as a genealogical descendant of COMIT suggest that the broadest context for Snobol's reception was in COMIT's association with MT research. "During the 1950s," as Peter Wegner put it, "it was felt that mechanical translation and other glamorous language understanding tasks could be greatly facilitated by the development of string manipulation languages with special purpose linguistic transformation aids" (Wegner 1985, 10). But the story of how Snobol came to be be used outside Bell Labs, and how its creators came to be stuck with their deliberately flippant choice of name, offers its own details.

Snobol Version History

In using the combined category "string and list processing languages" in her classification of programming languages up to 1969, Sammet recognized a common class of problems in symbol manipulation, marked by the necessity to allocate memory storage automatically, sometimes in the process of computation itself, rather than specifying memory needs definitively and in advance. Problems involving expanding and contracting lists of data items (of any type) and problems involving natural-language text commonly present this difficulty, as new nodes in a list, or new copies and combinations of string data, require memory to be allocated on the fly, while the memory previously allocated to discarded memory objects must be reclaimed.[6] At least in 1969, this common requirement certainly justified treating languages like IPL, Lisp, COMIT, and Snobol as a group.

Snobol 1 was preceded by the earliest appearance of so-called regular expressions, a standardized form of which is implemented by most imperative programming languages today (indeed, it is so widespread that "regexes" might be understood as a kind of embedded domain-specific language). Described by Stephen Cole Kleene in the early 1950s, regular expression syntax would become a characteristic feature of a group of frequently used text-processing utilities included in the Unix operating system, the best-known of which is the stream editor **sed**. Snobol's pattern-matching syntax is not to be conflated with regular expressions, however; while in discussing the former and proposing its integration into the latter, Jeffery and colleagues (2016) refer to "the ubiquitous regular expression notation introduced by [Stephen Cole] Kleene" (Jeffery et al. 2016, 1974), they are careful to distinguish them as the products of different designs.

Between 1962 and 1967, four major versions of the language were released, with Snobol 1 and Snobol 4 differing substantially enough from Snobol 2 and 3 to be considered by their creators to be separate languages (Ralph E. Griswold 1981, 601). According to Griswold, Farber first implemented Snobol 1 on an IBM 7090 in three weeks "early in 1963," and that implementation "was in extensive use within BTL" by the summer of that year (Ralph E. Griswold 1981, 603). Griswold describes the prompt embrace of Snobol throughout Bell Labs as "surprisingly enthusiastic," adding new applications in graph analysis, syntax analysis, text generation, and the simulation of automata to the PRSD's own usage, which had included evaluating algebraic expressions and a range of compilation tasks including both Fortran and assembly code generation. Public distribution began at the end of 1963, following enthusiastic responses to the announcement of Snobol 1 in public lectures given by Snobol's creators. In an early example of what would today be called the free and open-source software (FOSS) model, Bell Labs permitted the PRSD to distribute Snobol's source code and documentation without fees or restriction of use. Griswold would attribute a good portion of Snobol's success to this decision, contrasting it with the "tight control" exerted by MIT over the COMIT source program (Ralph E. Griswold 1981, 625–26).

"Users quickly outgrew" Snobol 1, Griswold reflected, "and the absence of certain facilities in SNOBOL made some programming tasks impractical or impossible." Applications in natural-language processing, not an area of focus of the PRSD and one in which Snobol adoption came as something of a surprise (at least to them), made the absence of a convenient way to measure the character length of a string a clear deficiency. A more general obstacle was presented by the unavailability of syntax for writing functions, modular units of reusable code that made programming significantly less labor-intensive. A revision that addressed some but not all of these issues, which was designated Snobol 2, was released internally at Bell Labs in 1964, but development continued at such a pace that a third version of the language had replaced Snobol 2 at Bell Labs by the end of the year. Snobol 3, which included a syntax for defining functions, was released publicly to an enthusiastic reception (Ralph E. Griswold 1981, 606, 619).

Despite Sammet's grouping of "string and list processing languages," it could be argued list- and string-processing languages represented distinct designs, intended for relatively distinct applications.[7] Snobol 3, Allen Forte remarked, "deals with a single basic data-structure: the free-form character string. By free-form character string is meant a character sequence of arbitrary structure, the elements of which are limited only by (machine-imposed) restrictions on the character set" (Forte 1968, 158). Three basic operations could be performed on instances of this central data type: naming (assigning a name to a string), concatenation (joining two or more strings together), and pattern matching (determining similarities and differences among two or more strings). Naming was performed using an assignment statement like s = "foo", a pseudo-algebraic idiom common to imperative programming languages then and still today. No operator was needed for either concatenation or pattern matching: the statement s = "foo" "bar" would assign to the name s the concatenation "foobar", while the expression "foo" "o" would succeed as a pattern match (because the character "o" is contained in the string "foo"), with a Boolean logical value representing true or false as its outcome.

The Boolean logical value produced by a pattern-matching operation could be captured by a goto expression, and this was Snobol's principal control flow feature: "foo" "o" /s(start) (in Snobol 3) or "foo" "o" :s(start) (in Snobol 4) would jump to the program line labeled start if the pattern-matching operation succeeded, while "foo" "a" /f(start) (in Snobol 3) or "foo" "a" :f(start) (in Snobol 4) would do the same if the pattern-matching operation failed.[8] Combined with the omission of operators for such operations as concatenation and pattern matching, which could be invoked by the simple juxtaposition of strings, this feature of Snobol's highly compact syntax made it both powerful, in that a great deal might be accomplished by a single short line of code, and difficult to read, for the same reason. By comparison, Snobol's syntax for functions was verbose and difficult to learn, and it had no other core syntax for either encapsulation or branching (conditional execution), so that extensive resort to what Edsger Dijkstra would denounce as the "go-to statement" was inevitable. Snobol's simple, direct input/output features, the power of its pattern-matching syntax, and its many other sophisticated and powerful affordances including partitioning, duplication, and replacement, dynamically named strings ("string variables"), and indirect referencing, in which the string value of a name could be repurposed as another name, made it easy, if often obscure, to do a great deal on a single line;[9] but executing portions of a written Snobol program in a nonlinear sequence was another matter.

```
loop s arb $ pre len(1) $ c len(1) $ d *lgt(c,d) rem $ post = pre d c
post :s(loop)
```

CODE EXAMPLE 5: Example of Snobol concision. This program sorts a string. Source: Paine (2010a).

Griswold and his collaborators considered Snobol 3 a mature string-processing language. In their view Snobol 4, publicly released in 1967, not only remedied perceived deficiencies in Snobol 3, but expanded it into a true general-purpose programming language, very substantially different from its precedessor. The most significant difference in Snobol 4's design was its introduction of nine built-in data types along with syntax for additional user-defined types, a radical departure from the single string type in Snobol 1, 2, and 3. One might imagine such a development the outcome of a lengthy interval of planning, if only informally, or at the very least of accumulation of evaluations, feature requests, and so on. But to hear Griswold tell it, Snobol 4, and all the effort that went into it, emerged almost by happenstance. As part of Bell Labs' participation in Project MAC (Project on Mathematics and Computation) at MIT, it proposed to replace its IBM 7094 computers with the 600 series machines supplied to MIT by General Electric (and used simultaneously in the development of the Dartmouth Time Sharing System). Faced with the need to reimplement Snobol 3 for the GE 645, Griswold and his collaborators produced a specification that departed so radically from Snobol 3 that for a time they considered new names for the language (Ralph E. Griswold 1981, 607).

Technically, this effort included Snobol 4 in the development of the Multics operating system, which emerged from Project MAC and would serve as the basis for the better-known Unix (Unics) OS after Bell Labs withdrew from the project in 1969—though Griswold noted that the structural independence of the PRSD persisted. "Although SNOBOL4 was officially part of the MULTICS project," he wrote, "the affiliation was largely ignored by both sides. No schedules were established, no goals were projected, and no reports of documentation were ever requested or supplied" (Ralph E. Griswold 1981, 607). As it happened, in the end the first release of Snobol 4 was for the IBM 7094 machines already in use at Bell Labs, and it was reimplemented for the IBM 360 when Bell Labs decided not to acquire the GE 645 after all (Ralph E. Griswold 1981, 608).

One might wonder why the innovations of Snobol 3 and Snobol 4, in particular the syntax for function definitions and user-defined data types, did not extend further into what would come to be called structured programming, the basic design concept of which (that is, the compound statement) dated to the appearance of Algol 58, a full decade preceding the publication of Dijkstra's polemic. The answer may be that four versions of a language in six years, at least two of them being significant revisions, is a great deal by itself, and that if Snobol 4 development had not ended with Griswold's move into academe and on to other projects, one might have expected such further development.

Snobol Today

Snobol 4 was well received. By 1976, nearly ten years after its first public release, Douglas W. Maurer would observe that "much of the computing world is now convinced that SNOBOL4 is not just a good programming language [...] it is an important advance in the state of the art of programming languages" (Maurer 1976, ix). Among reasons to suggest that Snobol design research remained active and to suppose that the developmental momentum generated in the mid- to late 1960s might last, one might cite Gimpel's (1972) commentary on an experimental implementation of block datatypes; the proposals for improving Snobol 4 pattern matching by Griswold's new colleagues at the University of Arizona, Frederick C. Druseikis and John N. Doyle, who complained of the "unbelievable" and "hopeless" complexities of extending the reference implementation's model (Druseikis and Doyle 1974, 311–12); Ralph E. Griswold's (1974) description of an experimental implementation called Snobol X, incorporating significant revision of control structures in Snobol 4; and Abrahams's (1974) lively but constructive rejoinder that such changes still did too little to control Snobol 4's "useless freedoms," made with the goal of encouraging further refinement. Yet there is, for example, Richard Dunn's "rather dismal appraisal" of Snobol 4 as permitting "too much freedom" for the task of bootstrapping a JANUS compiler in its use of "unusually high amounts of storage and time for rather simple tasks" (Dunn 1973, 31–32). In the end, no further development of Snobol 4 was undertaken at Bell Labs after Griswold's departure for Arizona in 1971, and the introduction of the personal computer over the course of the next decade would leave Snobol marooned—though not simply or straightforwardly so. In a memoir composed in 1985, Griswold reflected that Snobol 4 adoption declined through the mid- and late 1970s as universities and other research centers upgraded their facilities with systems for which Snobol was unavailable and would not be implemented. Whereas what one might call antiquarian and hobbyist implementations of Snobol 4 for PC operating systems gave the language a new lease on life, during this period, it was a tenuous one; Griswold is notably, perhaps irreconcilably, of two minds on the meaning of Snobol's rapid decline. "Even when SNOBOL4 became available for personal computers," he reflected, "it was hard to get the majority of the PC community to take notice. PC magazines were not interested and the word spread only slowly among specialized audiences and the 'underground'."

On the one hand, by 1985, when Griswold presented these remarks at ICEBOL 85, the 1985 International Conference on English Language and Literature Applications of SNOBOL and SPITBOL convened at Dakota State College, articles on Snobol 4 had reappeared in popular computing publications, with one describing it—in Griswold's paraphrase, at least—as the "exotic language of the decade." On the other hand, the experience of using on a modern PC any language still rooted in the era of batch-processed instructions punched on cards would inevitably be a somewhat alien, and alienating, experience. Articulating this structural contradiction for his audience of humanities scholars well versed in the language he had created, but effectively using it in its afterlife (an issue to which we return below), Griswold appealed to the genealogy of educational and expressive personal computing as it emerged in the late 1970s and early 1980s, when the popular Commodore PET, Tandy TRS-80, and

Apple II computers shipped with implementations of dialects of Dartmouth BASIC as their operating systems and/or user interfaces. Users of Snobol implementations on such early PCs, Griswold reflected, "have largely learned computing from scratch, re-experiencing the earlier era of computing in assembly language and using primitive programming languages like BASIC. Contrast BASIC and SNOBOL4! What a surprise it is to a person who has been limited to BASIC to experience the power and freedom of SNOBOL4!" (Ralph E. Griswold 1985, 4–5).

This suggests that Snobol's orthogonality, as Griswold understood it, at least (an understanding there is no good reason to challenge), was not limited to its prescience in treating strings as objects, in strong contrast with languages whose designs provided string operations and pattern matching only through extension, or which refused to recognize strings as a data type at all. Nor did it refer only to Snobol's more general design choices: its "typelessness" at a historical moment when strong typing was preferred, or the relaxed structure or nonstructure providing "ease of programming" and "freedom from constraints" at a time of "police-state" regulation of programming practice (almost certainly a reference to Dijkstra's polemics) (Ralph E. Griswold 1981, 635)—or to the assessments of Snobol's control structures as "unique," "awkward," "crude," "bizarre," or "degenerate" (Ralph E. Griswold 1974, 7–8). A third aspect of Snobol's orthogonality, as Griswold understood it, lay in its accessibility to consumer computing enthusiasts and other amateurs or nonprofessionals, a group which Griswold appears to have imagined including both professional humanities scholars and ordinary early adopters of the consumer PC. In being far more powerful than Basic, yet just as available and easy to use, Snobol established one additional and final oblique relationship to a computing "mainstream."

There is no question that in the case of Snobol, such relationships are relationships of decline, indeed at this point something close to disappearance. At the time of this writing, the only implementation of Snobol 4 (or of any of the Snobol languages, for that matter) that can be installed and used on a PC with minimal effort is Philip L. Budne's CSnobol4, a C language implementation with two major versions and eight releases spanning 2004–15.[10] I have successfully used CSnobol4 version 1.5, available through the Homebrew package manager and installer for MacOS operating systems, to write solutions for exercises in Allen Forte's 1967 *SNOBOL3 Primer: An Introduction to the Computer Programming Lan-*

guage and a 1971 textbook by Griswold, Polonsky, and J. F. Poage, *The SNOBOL4 Programming Language.*[11] DOS and Java implementations by Viktors Berstis and Dennis Heimbigner, by contrast, appear abandoned,[12] and while Catspaw, Inc. of Salida, Colorado, claims to provide Snobol 4 implementations for a variety of platforms, its website contains many dead links and its downloadable product order form displays the label "Prices [are] good through December 31, 2003."[13] It is very unusual to encounter a recent publication, either in software industry magazines or academic journals, that mentions any of the Snobol languages.[14]

Applications

COMIT's proposed applications in linguistics are best understood as an episode in the larger story of the displacement of historical linguistics, a humanistic endeavor, by its social-scientific successors, beginning with structuralism in its characteristic emphasis on synchronic (static), rather than diachronic (dynamic and historical) objects of investigation.[15] Some work in COMIT extended into literary study as well: Ryan (2017) mentions a project by Robert I. Binnick, research assistant in Yngve's lab at the University of Chicago, on a Proppian story generator written in COMIT.[16] But what appear in retrospect, here, as adumbrations of Snobol's late adoption in the humanities are perhaps not either simply or exactly that.

I have mentioned that according to Griswold, Snobol 1 quickly outgrew its initial use in algebraic formula manipulation, in 1963 alone adding Fortran and IPL-V code generation, Fortran compilation, and other applications in text generation, syntax and graph analysis, and simulation, all of this inside Bell Labs alone and prior to Snobol's first public release (Ralph E. Griswold 1981, 604). (Positioned at both the tail end of Snobol's development history and the extent of its range of applications was the purported interest of the US National Security Agency and Central Intelligence Agency in Snobol 4, which Griswold described with both pride and amusement.)[17]

While he grouped Snobol 3 with COMIT and Lisp as programming languages designed for nonnumeric computation (and as foundational to early research in artificial intelligence, in that respect), Forte had insisted on Snobol's distinction in enjoying a wider range of applications than either COMIT or Lisp, including not only "linguistics research" but "the construction of psychological models [. . .] cryptanalysis, music research, theorem proving, and parsing algebraic expressions" (Forte 1967, 158). The jacket copy of Forte's 1967 *SNOBOL3 Primer: An Introduction to the Computer Programming Language* described Snobol 3's applications in nearly identical terms, adding "the making of bibliographies and indexes." Rather pointedly, this editorial paratext identified the volume's author as "a humanist" (see below), a narrow designation incongruent with Forte's understanding of Snobol and a more accurate description of Forte's professional academic affiliations than of his research itself, with its deep and wide investments in musical applications of mathematics. (However much he may have served as a conduit for Snobol's adoption in the humanities, Forte was more ecumenical than the jacket copy of *SNOBOL3 Primer* suggests, in the sense that he appears to have tried out every computing solution available to him and that nothing special drew him to Snobol in the first place, though it turned out to serve his needs.[18] There is no reason to believe that either Forte himself or any other humanities scholar influenced the design and development of Snobol, at any time in its history.) Nearly a decade later, in *The Programmer's Introduction to SNOBOL,* Maurer would identify Snobol 4's principal uses in programming language implementation ("writing assemblers, compilers, interpreters, etc."), on the one hand, and "text processing in the humanities and social sciences," on the other. Still, it was the former, not the latter, that Maurer emphasized: "From a practical standpoint its utility in systems programming, particularly from the standpoint of reducing programming time, is little short of miraculous. Assemblers that used to take six man-months can now be written by one man in five weeks." And Maurer went out of his way to recommend Snobol 4 as a general-purpose programming language, diffusing its origins considerably: "SNOBOL is more than a string-processing language; it is an algebraic language, a pattern-matching language, and an associative language, as well. In addition, SNOBOL gives rise to two new *ways* to program—associative programming and pattern programming" (Maurer 1976, ix–x). In contrast, Peter R. Newsted inverted this approach, recommending Snobol for simulations, game programming, and compiler writing, but emphasizing natural linguistic research applications including text searching, concordance generation, syntax and sentiment analysis, transformational grammar and phonological rule testing, and content analysis (Newsted 1975, 10–13).

This variety of domains and use cases, miscellaneous or universal as the case may be, makes the story of Snobol's origins a difficult one to tell plainly. What is clear is that in the long run, Snobol did not survive in its projected role as a general-purpose programming language. Indeed, it never flourished as one. Ruth H. Sanders's retrospective evaluation is most apposite: "SNOBOL is a general-purpose programming language and can be used to solve most problems which are considered solvable by computer. However, it is particularly well suited to manipulating language. It has been used in many text manipulation programs in linguistics, social science, and library science" (Sanders 1985, 45). Still, by the 1960s not only social scientists, but humanities scholars with research computing needs were already actively using Fortran, Algol, PL/I, APL, and Basic, among other major programming languages, as documented in the pages of the journal *Computers and the Humanities*. Well aware of the availability of Snobol 3, Heller and Logemann (1966) had instead promoted PL/I for humanities research beginning in 1966. Subsequently, Michael P. Barnett's SNAP was described by its creator as a programming language designed explicitly for humanists, rather than merely adopted by or adapted for them (Barnett 1970). A year later, Raskin (1971) described Snobol 4's string-processing features as making "both ALGOL and FORTRAN look clumsy," but noted that Snobol took no less time to master, was more difficult to read than Fortran or Algol and more difficult to debug owing to the limitations of the tools available, and was rivaled by APL for text-processing applications despite the latter's idiosyncrasies. In the end, Raskin recommended that "a beginner given a choice of ALGOL, FORTRAN, and PL/I should probably choose PL/I," notwithstanding Snobol 4's string-processing "elegance."

On the whole it would appear that as Snobol's range of domains and use cases narrowed, during its decline, its applications in social-scientific and humanistic research—present almost from the beginning, yet nondeterminative, possibly even arbitrary—were precipitated out, providing it with its remaining life as a programming language in active use. Diffusing somewhat the normal sense of the term, one might therefore imagine Snobol's place in the history of programming languages as the place of a specific, unusual type of domain-specific language. Not, that is, according to the normal denotation of that term, a specialized rather than general-purpose language adapted to a specialized, specifically technical task: structuring web pages, for example, or synthesizing music, or describing an integrated circuit. Rather, as a programming language that is neither of general purpose (productively usable in potentially any domain) nor domain-specific (productively usable in only one), but instead what we might call domain-*in*specific, in the sense that its domain never became clear and it kept drifting. Griswold's own reflections on this point are clear enough. "The original objectives were greatly exceeded," he wrote,

> but the areas of application for SNOBOL4 [. . .] are substantially different from those originally anticipated. For example, during the development of SNOBOL, the designers gave no thought to potential uses for document formatting or for computing in the humanities, although these have become some of the most significant applications of SNOBOL4. While SNOBOL has had some applications in formula manipulation, the problem that provided the initial motivation for the development of the new language, it has never enjoyed significant popularity in this area. Similarly, use of SNOBOL as a compiler-writing tool was envisioned, but the suitability of SNOBOL4 in this area is at best controversial. In fact SNOBOL4 is not now widely used for implementing production compilers. (Ralph E. Griswold 1981, 625)

This is not to deny that Snobol became a "humanities programming language" at the end of its life in more or less the sense telegraphed by the jacket copy of Forte's *SNOBOL3 Primer,* identifying its author as a "humanist" (as noted above) and "Professor of Music at the Massachusetts Institute of Technology." The jacket copy advertised Snobol in terms that readily recall Yngve's early briefs for COMIT, insisting that Snobol "is easily learned by novices at programming" and "especially useful to those in the humanities and social sciences who could make use of a computer in their research." "Professor Forte, himself a humanist," it concluded, "develops the language completely, taking an informal and conversational approach" (Forte 1968). Forte's preface took this point further, constructing an active frustration among scholars in nontechnical disciplines. "Many scholars who would like to utilize computing machines in connection with research," he began, "are discouraged by the formidable lingua franca of the computer world." Forte described his method as assuming "no previous experience with computing machines" and "no prior qualifications whatever for computer programming," promising that "the experienced programmer reads this primer at his own risk" (Forte 1968, v).

Not without good reason—and yet not without any reason to hesitate, either—the humanist and the novice, inexpert, or casual programmer were often conflated, in this late Snobol evangelism. Griswold identified the humanities as an academic domain "where sophisticated programming skills are less common" (Ralph E. Griswold 1981, 626), and according to Timothy Montler, the frustrations of such professionals as described by Forte persisted nearly two decades later. "Keep in mind," Montler wrote in his contribution to the proceedings of ICEBOL 85,

> that none of the people involved in the projects discussed here is a professional programmer. They are professional linguists and lexicographers who want to be able to devote their attention to the complex natural languages they study. They have each experienced the frustration of consulting with a computer professional, not understanding what he is talking about and knowing that he does not understand them. Most commonly and most frustratingly the confident computer person will try to convince them that all they need is a good word-processor [. . .] In teaching SNOBOL, I have often found that complete computer novices, mathophobes, and those of more poetic temperament produce better, more SNOBOL-like programs, than experienced computer science students. (Montler 1985, 181)

The repulsion of "those of more poetic temperament," and its redress by Snobol imagined as a "right-brain language" (Sanders 1985), are marked by the lines of Archibald MacLeish's "Mother Goose's Garland" used as string examples throughout Farber, Griswold, and Polonsky (1964), and in Griswold's explicit reflections as well. "I would personally like to think," Griswold remarked, that "SNOBOL had some responsibility in opening the use of computers in areas that had previously been nontraditional: areas in the humanities—music theory, among the early uses" (Ralph E. Griswold 1981, 655). In the last published Snobol primer for novice programmers, issued by Oxford University Press in 1985, Susan M. Hockey omitted the rougher textures of Snobol's provenance and domains of application, leading with what was effectively its afterlife. "Computers evolved from mathematical studies," she wrote, "and most programming languages were designed to solve mathematical or scientific problems or to process numeric data. SNOBOL, on the other hand, was written to handle text rather than numbers and is therefore particularly suitable for studies in the humanities" (Hockey 1985, v). While there is nothing especially objectionable in this formulation, particularly in its purpose and context (a concise instructional textbook), it does force us to recall that the text-processing problems for which Snobol 1 was devised were problems in applied mathematics and the technical sciences, not in the humanities, and that Snobol's designers had rejected COMIT precisely *because* of its orientation toward research on natural language.

The proceedings of ICEBOL 85, published the same year as Hockey's *Snobol Programming for the Humanities,* included papers on applications of Snobol in analyzing the poetry of Robert Frost and the programming of "computable poems" as well as in Northwest US Indigenous or Native American language data processing. Also included were papers on topics in nonliterary text analysis and on more technical issues including phototypesetting, interactive debugging, and one of Snobol's successors, the Icon programming language. Tosh (1985) described Snobol quiz and tutorial programs for studying the German language, transformational grammar notation, and phoneme comparison; analysis programs designed for work with Old and Middle English text files; programs serving as simulations of episodes in linguistic history (for example, the "great vowel shift" in English) or of texts like the I Ching; and student programs written in Snobol including text reformatting routines, word frequency list and keyword-in-context index generators, and an attempt at multilingual word-for-word machine translation. Montler (1985) described Snobol use in creating dictionaries, grammars, and collections of texts in the Salishan family of Native American languages. Raabe (1985) promised "decreasing reliance on guesswork as we attempt to understand and express the complexity of the creation of literature," while Strange (1985) opined excitably that "deconstructionism," then a leading analytic methodology in literary research that was furiously resented by its competitors, "looks more and more like a desperate case of computer phobia" (207).

Legacies

I have already mentioned Griswold's contribution to the ICEBOL 85 proceedings, titled "SNOBOL: A Personal Perspective." While suggesting that a brief Snobol revival was under way, spurred by antiquarian and hobbyist implementations for the new personal computers, the paper's mood was subdued, even elegiac. There is little to suggest that Snobol was widely used in the humanities and the qualitative social sciences after this point in the mid- to late 1980s, let alone anywhere else. Neither of Snobol's immediate successors, the Icon and Unicon programming languages (the former designed by Griswold himself), appears to have been anywhere as widely used as the parent language may have been at its peak. In the late 1980s Icon, carrying forward Snobol's orthogonal attentiveness to string processing as the "step-child of computation," was once imagined as an inheritor of Snobol's applications in the humanities and an exciting new medium for both humanities computing and beginning or casual programming instruction (Olsen 1987, 61), but an OCR "rescue" edition privately prepared in 2010 from the first edition of Alan D. Corré's *Icon Programming for Humanists,* originally published by Prentice-Hall in 1992, notes that "for a long time it has been almost impossible to obtain a copy [of this book]" (Corré 2010, ix).

Set alongside the self-serious and rather technocratic literary-critical speculations of Raabe and Strange, both literary scholars with faculty appointments in English departments, the humor and self-deprecation, indeed the antidisciplinary energy of Griswold and his colleagues are both more memorable and more interesting in their intersection with questions of technical expertise. Snobol's deliberately "embarrassing," indeed "disgusting" name had been the subject of much variation, which saw associated tools and modifications or dialects of the language baptized ICEBOL, SPITBOL, FASBOL, SLOBOL, ELFBOL, SNOBAT, and SNOFLAKE, among other names. As disciplined as it mostly was, aspects of Snobol's implementation could inject levity into the user's experience as well. "The relatively lame humor exhibited in the choice of SNOBOL," Griswold recalled, "was not limited to the name of the language itself": Snobol error messages composed by Farber supposedly included "ALL OUT OF SPACE, YELL FOR HELP," referring to memory usage, and "NO END STATEMENT, WHISPER FOR HELP," reporting a syntax error (Ralph E. Griswold 1981, 605).

In some ways, Snobol's culture of humor may represent the most vivid trace of its history today. In an essay titled "Programs That Transform Their Own Source Code; or: the Snobol Foot Joke," Paine (2010b) noted the extreme inconvenience of Snobol's syntax for defining functions, but also the ingenious ameliorations created by Snobol programmers, including the runtime code modification, generation, and execution made possible by Snobol's sophisticated pattern matching. In describing the power and flexibility of the latter—impressive enough even today that "discovering the power hidden in this ancient language is like winning a prize" (Paine 2010a)—Paine appealed to a well-known trope and figure of programming lore, the so-called "foot gun," named for the act of shooting oneself in the foot.

In all likelihood the idea that the C programming language is a foot gun, meaning that it offers great power at great risk and that programmers using it often make potentially catastrophic mistakes, dates to its spread in the 1970s. Templeton's (1995) collection of humorous texts that had circulated on the Usenet forum "rec.humor.funny" during the 1980s included a chapter titled "Computer & Science Jokes." Among the texts therein is a three-part sequence with the titles "What Language Is Afoot?," "More Prog. Lang. Help," and "Shoot Yourself in the Foot Yet Again," each taking the form of a list of programming languages accompanied by humorous annotations and attributed to the Usenet identities of Brian Pane of the University of Florida, R. A. Montante of Indiana University, and Paul L. Schauble, respectively. The first of the three, "What Language Is Afoot?," began by describing use of the C language, canonically, as follows: "You shoot yourself in the foot." Each of the following entries for other programming languages worked some variation on this theme, from "You accidentally create a dozen instances of yourself and shoot them all in the foot" (C++, referring to its addition of object-oriented programming of C) to "After realizing that you can't actually accomplish anything in the language, you shoot yourself in the head" (Modula/2) to "You hear a gunshot, and there's a hole in your foot, but you don't remember enough linear algebra to understand what the hell happened" (APL) (Templeton 1995, 9). "More Prog. Lang. Help" added Algol ("You shoot yourself in the foot with a musket. The musket is aesthetically fascinating . . .") and Basic ("Shoot self in foot with water pistol"), among others. "Shoot Yourself in the Foot Yet Again" contributed annotations for Prolog ("You attempt to shoot yourself in the foot, but the bullet, failing to find its mark, backtracks into the gun which then explodes in your face"), Forth ("yourself foot shoot"), and assembly language ("for those who like to load their own rounds before shooting themselves in the foot") (Templeton 1995, 10–11).

In a webpage titled "How to Shoot Yourself In the Foot," Stepney (n.d.) has continued collecting foot gun jokes, including versions for Pascal ("The compiler won't let you shoot yourself in the foot") and Java

("You locate the Gun class, but discover that the Bullet class is abstract, so you extend it . . ."), among others. Arguably, the humor of such self-deprecation is effective because it captures both the apparently intractable difficulty of computer programming, as a human cognitive activity, and the specific hazards of software engineering as an occupation and a profession that after nearly seventy years has yet to develop standardization and certification practices comparable to the other engineering professions.[19] The foot gun joke for Snobol captures quite well the extreme flexibility that arguably makes Snobol special, and at the same time merely an extreme illustration of the opportunities and hazards made possible by treating program code as just one more interpretation of data. As included in both Templeton (1995) and Stepney (n.d.), the entry for Snobol in "More Prog. Lang. Help" is as follows: "You grab your foot with your hand, then rewrite your hand to be a bullet. The act of shooting the original foot then changes your hand/bullet into yet another foot (a left foot)" (Templeton 1995, 10). Note that in this scenario, there is no concrete object resembling a gun, and nothing is explicitly named "gun." The foot, which in other jokes is (obviously) the gun's target, and an emblem of the human agent and beneficiary of the magic of programming, is *also* the gun—more precisely, no distinction is ever established between them. Upon being gripped by the hand, the foot takes the role of a gun (without, note again, ever being named as such), upon which the hand is "rewritten," changing it into another object, a bullet. "Shooting the foot," here, is not, as in the other jokes, impacting the human foot with a bullet fired from a gun. Rather, it means "firing the foot-as-gun." Upon this foot-as-gun being fired, the bullet that was recently a hand is in turn rewritten, as another foot. One senses that the Snobol foot joke might be extended indefinitely, in a simulacrum of the recursive automation that is a defining characteristic of the programming language hierarchy, from machine code to assembly mnemonics and higher-level languages and onward to fully automated code generation and programming by machine learning.

Thus, where C is a foot gun, Snobol is neither and both at once, which is to suggest that its promise and danger is of another order. Of the factors that express this difference, the easiest to grasp is probably string handling. Though Ritchie (1996) makes no mention of Snobol, both languages emerged from Bell Laboratories, with the first versions of C dating to a few years after the final release of Snobol 4. Still, the two languages' implementations of string handling could not have diverged more. Whereas Snobol implemented strings as what we now call objects, replete with their own properties and behaviors, C implemented strings as null-terminated arrays of characters, and where string operations were core elements of Snobol, usable without any extra ceremony, C segregated string operations in standard but external library routines, which had to be explicitly loaded.

Ritchie (1996) had little to say about C string handling except to note that "none of BCPL, B [languages that served as models for C's design], or C supports character data strongly in the language" (Ritchie 1996, 675). While admitting that "certain string operations are more expensive than in other designs," that "the burden of storage management for strings falls more heavily on the user," and that "C's treatment of arrays in general (not just strings) has unfortunate implications both for optimization and for future extensions," Ritchie insisted that "C's approach to strings works well" (Ritchie 1996, 675, 684–85). The implementation of strings as null-terminated character arrays is nonetheless widely regarded as a source of agony in programming in C. C's successor, Bjarne Stroustrup's C++, initally repeated the mistake, which Stroustrup acknowledges was among the worst defects of C++ 1.0 (Stroustrup 1996, 751); but it was eventually remedied in (what is now) the conventional way, by adding a standard string class as well as wrappers for C character arrays. Since Sun Microsystems' Java language formalized this correction of string handling (among much else) by setting aside everything C++ had retained to provide compatibility with C, no major language designer has ever looked back.

To be sure, Ritchie had his reasons for designing C as he did, and there is nothing wrong with the implication of his omission: that Snobol and C were, so to speak, apples and oranges. C is a high-level language for the manual management of memory, a task and a problem that Snobol set aside more or less entirely, in the first place, in focusing on string data. It is in their incommensurability that Snobol and C might be understood as mutually reflected images or "rememories" of each other, emerging as they did from the same institution at virtually the same moment, at the dawn of the modern era of programming. They are as different as can be, but of course that is in no way absolute. Snobol may have presciently treated strings as objects, in strong contrast with C's refusal to recognize them as a data type at all; but in other ways, as I have indicated, Snobol's syntactic compression and absence of structure invite the descriptor "high-level assembly" in its own, quite different way. As Griswold put it, in his remarks on Snobol's orthogonality, "It not only has GOTOs, it has side effects, it has all of the lovely things of assembly language, everything that goes with it" (Ralph E. Griswold 1981, 656).

This is not to be either plainly or simply lamented. We might even say that if we have more or less clearly reached the end of the line, so to speak, in this particular genealogy of making and building with these our highest-level tools and machines, which we so misleadingly call "languages," that is much the better for history?

———

In this chapter, we have performed our first close study of a specific programming language. In the following chapter, we turn to a second specific programming language, JavaScript, which appeared shortly after the Snobol languages fell into obscurity and would develop into the most widely used programming language in the world today, a history that rather neatly inverts Snobol's evolution from a project of and for computing experts and specialists into a utility for humanities scholars. Though unlike the Snobol languages, JavaScript's technological-historical role as a mediator of the relationship of professional technical expertise to casual, inexpert, and nontechnical uses of computing was imagined in its original design, the historical development of JavaScript also inverted its own original purpose. Together, then, Snobol and JavaScript constitute the two parts of a dual case study, whose contrasts illuminate the complexity of the links among software, automation, and expertise.

FIVE

JavaScript Affogato

NEGOTIATIONS OF EXPERTISE

Transitions

Two thousand sixteen was an inconspicuously transitional year for the information space once commonly referred to as the World Wide Web (WWW). Those attentive to linguistic usage will recall that the 2016 edition of *The Associated Press Stylebook and Briefing on Media Law* released in June recommended that the words "internet" and "web" no longer be written with initial capital letters (Alba 2016), in a sign that the propriety marked by their referents' novelty had finally settled, or worn off. For those more attuned to matters of technical infrastructure, what may come to mind instead is the announcement by Oracle Corporation that its Java web browser plug-in would be deprecated in the forthcoming ninth version of its Java Development Kit (JDK), a platform for writing and packaging software applications in the Java programming language (Warren 2016). Taking these two real, if lesser milestones together, it seems safe to say that for anyone who remembers the original promise made for Java applets as a common WWW technology, at their moment of emergence in the mid-1990s, this was a chapter of recent technological and cultural history quietly coming to an end.[1]

To be sure, Oracle's hand had been forced by Microsoft, Google, and Apple, which had either reduced Java plug-in support in their browser products or removed it entirely. And yet embedded Java applets had long since become a legacy technology, still useful for some computationally intensive graphical visualization tasks (disproportionately in scientific applications), but no longer in wide use outside that domain. Whether or not they are old enough to remember the role originally imagined for Java in the browser, in particular, most of those who design websites and program web applications for a living today would be unlikely to regret their eclipse. Even before the emergence of personal data security as a substantive public issue in 2012, Java applets presented grave, often intractable security risks that web developers had had to learn how to manage, or ignore. A more general reason for the irrelevance of Java browser applets by 2016 was a historical one, linked to changes in the profile of the Java programming language and Java programmers in the software development industry as a whole. When Fredrick P. Brooks Jr. chose for the seventh chapter of *The Mythical Man-Month: Essays on Software Engineering* (1975) the title "Why Did the Tower of Babel Fail?" he was reflecting on the biblical story of Babel as a fable of engineering (the hubristic or merely presumptuous construction of a tower tall enough to reach heaven), rather than a fable of language (divine punishment imposed in the form of linguistic difference and permanently impaired communication). Nevertheless, *The Mythical Man-Month*, the first widely read and still the most celebrated reflection on managing large software projects, was also an informal study of communication, not excluding the metaphorized communication that a software programmer struggles to achieve with a machine.[2]

This chapter is a study of the negotiation of technical expertise, specifically the technical expertise involved in software programming, and in particular that involved in programming websites and applications— that is, what is today called "web development." In this case I take my bearings from the present historical moment, understood as a long interval of economic constraint beginning in 2008 and shaped by both economic and political investment in "coding" instruction as job retraining for unemployed and underemployed US blue- and white-collar workers alike.[3]

Java and JavaScript

In December 1995, when Sun Microsystems and Netscape Communications issued a joint press release announcing "JavaScript, the Open, Cross-Platform Object Scripting Language for Enterprise Networks and the Internet" ("Netscape and Sun Announce . . ." 1995), Sun's Java programming language was already well on its way to achieving the virtually uncontested market dominance, comparative prestige, and privilege as an instructional language that it would enjoy for a decade and more. Though Java 1.0, the first public release, had appeared only the same year, Sun's promise of true platform-neutrality and portability for the Java Runtime Environment was immediately attractive to enterprise software developers tiring of the demands placed on them by the C and C++ languages then widely in use. Java promised to moderate some of the complexity entailed by the access both C and C++ provided to low-level memory management, as well as the specific complexities introduced by C++ imagined as "C with classes,"[4] without reducing the power and expressivity those languages offered to enterprise systems programmers specifically. Though it was initially designed for the lightweight hardware application of embedding in programmable consumer appliances, and only later adapted for serving and embedding in HTML pages, Java was very much a professional's language, restrictive in its requirements for data types (being both statically and strongly typed) and in its promotion of a single programming style, the object-oriented programming (OOP) paradigm it would help popularize, as well as in the verbosity that both these forms of restriction produced. Presented as a professional alternative to both C and C++ rather than a radical departure from either, Java's relative ease of use included no special claims of approachability for inexperienced coders or nonprofessionals.

JavaScript was different. Sun and Netscape's press release used the word "complementary" three times to describe JavaScript's relation to Java: "JavaScript as a complement to Java" (in the document's subtitle); "The JavaScript language complements Java"; "[JavaScript is] complementary to and integrated with Java." Java, the press release emphasized, "is used by programmers to create new objects and applets," while JavaScript "is designed for use by HTML page authors and enterprise application developers to dynamically script the behavior of objects running on either the client or the server." If the mention of "enterprise application developers" and server-side applications suggested a place for JavaScript in the established industry of Java development, the sentence that followed better illuminates how HTML page authors were imagined, and how they imagined JavaScript's complementing of Java in a different sense. "JavaScript is analogous to Visual Basic," it read, "in that it can be used by people with little or no programming experience to quickly construct complex applications" ("Netscape and Sun Announce . . ." 1995).

From its introduction, JavaScript's reception by software developers, and its importance in "web development" as we now understand it (as an area of either software development or graphic design, depending on whom one asks), was structured by a continuous negotiation of expertise. Especially today, it is rare to encounter an introductory tutorial or textbook for beginners that fails to pause to disambiguate JavaScript from Java before undertaking to cover even the basics.[5] Most often, and especially today, the motive for such disambiguation is less to clarify the historical relationship of these two languages than to clear a space for JavaScript by separating it from association with Java—specifically, with Java's verbosity and its object-oriented programming paradigm, and perhaps from Java's association with enterprise application programming, the drudge work of software engineering—and its diminished presence in the more flexible and experimental startup culture of the 2000s and 2010s, as well. While such gestures are understandable at a moment when Java's reputation is more or less clearly in decline,[6] they can obscure the historical entwinement of these two languages, with consequences that are regrettable from any but the most purely practical or instrumental perspective.

I use the term "improvised expertise" to describe both conditions for and effects of the unanticipated development of JavaScript from a mere complement to Java, designed for casual and inexpert programmers, into a language whose range and complexity of use has now propelled it ahead of Java in some ways, even (by some measures, in some domains) where Java once dominated. My argument is that such "improvised expertise" separates JavaScript at least partly from other, otherwise similar experiments in making programming accessible to nonexperts, from the original BASIC language, developed as an instructional language at Dartmouth College in the 1960s, onward.

The concept of improvised expertise also encapsulates the conditions for and effects of three specific developmental dynamics in JavaScript's recent history. First of these is a rapid acceleration in development of the programming language itself, now occurring at such a pace that ECMAScript, the specification on which JavaScript is based, shifted in 2015 from using traditional ordinal version numbers for editions to a year-based designation (so that the official name of ECMAScript version 6 became ECMAScript 2015, with new editions released on a yearly schedule from that point forward). Second is the abrupt emergence and extremely rapid growth of JavaScript in server-side networking, data processing, and other so-called back-end development tasks, a domain traditionally handled separately from the user-facing, design-oriented front-end site development that Sun and Netscape's 1995 press release suggested would be JavaScript's main use case. Third is the equally relatively recent and abrupt, yet decisive emergence of JavaScript as the dominant language of a new generation of dynamic web application frameworks (principally Ember.js, AngularJS, and Facebook's React, but also Meteor, Express, and others) and the developer tooling suites that support them, in a partial displacement of the Ruby language–based Rails framework popularized during the late 2000s.

This rapid, largely unanticipated growth in JavaScript's range of application and its general importance in the software industry has even seen it enter elementary computer science instruction as language of preference, in some cases displacing Python (which itself has selectively displaced Java) in the classroom. Here, the phrase "improvised expertise" marks a paradox: while core JavaScript remains a small, approachable programming language when abstracted from its main domains of application, website and application development, using JavaScript professionally in those domains today is virtually impossible without very substantial, ongoing study of the language's advanced features and support for multiple programming paradigms, as well as of the new JavaScript-based development frameworks and tooling suites, the frenetic development pace of which virtually ensures that they will be replaced by other, newer frameworks and tools before they emerge from beta status and a commensurate level of documentation. This means that the learning curve for new professional JavaScript developers—not to mention the nonprogrammers JavaScript was originally designed to serve—will be very steep indeed, and it suggests that sooner or later, JavaScript's improvised expertise will have some part to play in the disappointments of the latest push for universal programming instruction and other economic management schemes that conflate programming skills with basic literacy (that is, reading and writing in human languages) and with basic so-called computer literacy, as well (that is, using both general and domain-specific prepackaged software applications effectively). Where JavaScript's history as a programming language is in many ways a routine, if interesting case of simplification producing complexity, the logic by which everyone learns to code, and its variants, are arguably repetitions of magical thinking about the management of complexity in software production itself, with these two dynamics converging in the historical present.[7] In that sense, what we call "JavaScript" is not just a programming language, and not just a collection of environments and tooling supporting a programming language, including specifications and other documentation, implementations, and primary and secondary program artifacts (from development tools and frameworks to specific interpreters or "engines," compilers and transpilers, and other software components embedded in a browser or server applications). JavaScript can, at least at the moment and for the near term, be understood also as an assembly of broader technical and technical-historical dynamics, labor and management practices and arrangements, and discourses about education, job training, and production that privilege technical expertise, but also seek to generalize it in and for a demarcatable historical interval.

System and Scripting Languages

The first edition of ECMA-262 (ISO/IEC 16262), Ecma International's specification for ECMAScript, a standard for JavaScript, was published in June 1997. Edited by Guy L. Steele Jr., it described ECMAScript as a scripting language, defined as "a programming language that is used to manipulate, customize, and automate the facilities of an existing system" ("ECMAScript: A General Purpose, Cross-Platform Programming Language. Standard Ecma-262, June 1997" 1997, 1), rather than being used to create a new system. It acknowledged that the "existing system" of ECMAScript's original design was a World Wide Web page browser and a Web-based client-server architecture more generally, but also insisted that the ECMAScript specification had been written with a variety of possible host environments in mind ("ECMAScript: A General Purpose, Cross-Platform Programming Language . . ." 1997, 2). The first edition of ECMA-262 was equally pointed, and in some ways more specific, in emphasizing that "a scripting language is intended for use by both professional and non-professional programmers, and therefore there may be a number of informalities built into the language" ("ECMAScript: A General Purpose, Cross-Platform Programming Language . . ." 1997, 1). The history of what we now call higher-level programming languages is of course a history of efforts to make programming less arduous for *professional* programmers, as operation codes provided mnemonics for instructions that could otherwise only be expressed in binary, octal, or other numeric form, followed by what we now call programming languages providing another, platform-independent layer atop the hardware-specific operation codes, a layer still more remote from numeric encoding and apparently closer to natural language (then and still today, the English language specifically).

Efforts to make programming accessible to nonprofessionals did not, as one might expect, lag behind the effort to make programming more convenient for professionals; rather, they were coterminous and developed in parallel, not without significant overlap. At Dartmouth College, John Kemeny had devised DARSIMCO (DARtmouth SIMplified COde), "Dartmouth's first crack at a simple computer language" (Kurtz 1981, 516), a year before the appearance in 1957 of Fortran, the first widely adopted and lasting example of a higher-level or "third generation" language.[8] "Dartmouth students," Thomas E. Kurtz recalled in 1978, "are interested mainly in subjects outside the sciences," and most of the future "decision makers of business and government" among them were not science students (Kurtz 1981, 518). The rationale for Dartmouth BASIC, or "Beginner's All-Purpose Symbolic Instruction Code," was to provide such students with experience in writing programs (rather than merely learning about computer use) without having to understand operation codes "or even FORTRAN or ALGOL" (the latter another higher-level language developed in the 1950s) which Kurtz and his colleagues considered "clearly out of the question. The majority would balk at the seemingly pointless detail" (Kurtz 1981, 518). But the growth of Dartmouth BASIC into an entire family or class of languages represented its dissemination not only as an instructional language, but in some lines of development (such as that which produced Microsoft's Visual Basic) as "a production programming language for professionals" as well (Kurtz 1981, 547).

Today, terminological usage more or less clearly distinguishes "scripting" languages from "system programming" languages. System programming languages like C and C++ were designed to abstract away much of the detail of assembly-language programming (that is, programming in operation codes) while still leaving the programmer facilities for manually allocating and deallocating memory and thus staying "close to the metal," as programmers like to say, while enjoying the benefits of higher-level abstraction where that was preferred (for example, in syntax for iteration, branching and other control structures, function calls, and creating and managing collections of items of data). Managing memory efficiently involves distinguishing clearly among different data types (primarily, between mathematical and textual data types) as one makes use of them, so that no more memory than is needed is allocated for storing an item of data, and compilers for system programming languages typically enforce such discipline in the programmer—for example, by refusing to compile a working executable otherwise. Scripting languages, by contrast, abstract away and automate both data typing and memory allocation and deallocation, for the convenience of the programmer. This is partly because they they can take for granted the presence of an underlying system programming language and its libraries, for whose components they serve as a kind of adhesive or connective tissue, and in which they themselves are implemented (that is, the interpreter that provides a scripting language with its execution environment is itself a system-level language program).

Since the 1990s, however, various factors including the accelerating sophistication of hardware and innovations in programming language design have eroded some difference in the performance of scripting languages relative to system programming languages, at least in specific environments and for specific applications, and significant gains have arguably been made in some measures of programmer productivity. The economy of expression made possible once memory allocation and data typing are abstracted away can be fairly dramatic. If code in a system programming language like C is three to six times shorter, in countable individual instructions, than its equivalent in assembly language code (Ousterhout 1998, 24), the same instructions in a scripting language like Python might be half as long as their equivalent in C, C++, or Java syntax, and depending on the task possibly much shorter than that.

Other Contexts

Although it took nearly two decades, it was JavaScript, rather than Java itself, that made good on the promise of programmable Web pages and the browser application as a distributed multi-platform environment. As HTML-based Web publication promised to disrupt local monopolies of print publishers, JavaScript promised inexpert programmers access to a scriptable environment, while Java was to do the heavier lifting. In one of many interesting early formulations, the Web was imagined as a "shell" for interactive application development, by analogy with the AI-oriented "expert system" shells developed for use with Lisp and Prolog and marketed for rapid application prototyping in Java and other languages.[9] But the popularity of the Web was also used to justify the teaching of JavaScript to novices and as a "precursor to Java."[10]

The fading of Java's promise as a browser language did not immediately elevate JavaScript. One writer of the late 1990s correctly anticipated the development of browser-independent implementations of JavaScript (fully realized in 2009 with Node.js, discussed below), but incorrectly expected JavaScript to be displaced by Perl as a browser scripting language.[11] Today, after twenty years of emphasis on JavaScript's role in client-side web development (that is, on the software browser's presentation of data to the user), it is seldom remembered that Netscape Communications had explored server-side applications for JavaScript from the start. This is clear from the language of the 1995 joint press release with Sun, which specified that "JavaScript is an easy-to-use object scripting language designed for creating live online applications that link together objects and resources *on both clients and servers*," and that it was "designed for use by HTML page authors and enterprise application developers to dynamically script the behavior of objects running on *either the client or the server*" ("Netscape and Sun Announce" 1995, emphasis added).[12]

Still, it is not difficult to identify in retrospect some conditions that arguably later served JavaScript's explosive growth, including developments virtually coterminous with its first appearance. On April 30, 1995, the US National Science Foundation's NSFNET, a publicly funded network of supercomputer centers and telecommunications backbones serving academic research, was decommissioned, and the Internet as we know it today, unthinkable without private telecommunications carriers and Web-facilitated "e-commerce" and "B2B" or business-to-business transaction activity (to use two terms common in the mid- to late 1990s), began to take shape. America Online and Prodigy, up to that point private "online service" providers, also began offering access to the open Web. When the "Guide to the World Wide Web" created by Stanford graduate students Jerry Yang and David Filo was rebaptized "Yahoo!" and acquired the yahoo.com Web domain, large-scale Web indexing as a service was born.

Financial speculation linked to all these developments drove the Dow Jones Industrial Average past the 4,000-point threshold in February 1995 and the 5,000-point threshold in November, making two historic transitions in a single year. In this context, we are justified in remarking the larger context of the moment when JavaScript emerged as an instance of what I am calling "improvised expertise." Facilitated by new, consumer-friendly electronic financial networks and services, so-called day trading by individual small investors would grow by the late 1990s into a widely publicized pastime. Day traders responded rapidly to intraday price movements and sought out (as well as exacerbated) price volatility, buying and holding stocks for as little as a few minutes at a time and making a point of closing their positions at the end of each day. Commercial service centers opened to provide such traders with the network and PC hardware, software, and data and financial services then unavailable to home PC users. As a mode of improvised expertise permitting individual, often inexperienced and inexpert speculators to bypass both the authority and the fees of stockbroker and other expert (or at least certified) financial service providers, day trading was associated with the volatility of so-called Internet stocks and the improvised company creation and management practices of the dot-com bubble, and it was famous for the financial disasters such securities inflicted on day traders themselves, long before they triggered a US economic recession.[13]

The opening of securities markets to a new class of investor whose expertise was improvised, at best, was not the only significant economic event of 1995 and the years following it. It was in February 1995 that the 233-year-old Barings Bank, one of the world's oldest surviving financial institutions, collapsed due to losses incurred by a single Singapore-based derivatives trader who relied on the global distribution of Barings's operations help him evade scrutiny of his activities. Billionaire business publishing executive Steve Forbes launched his campaign for the 1996

Republican presidential nomination, refusing matching funds from the US Federal Election Commission to avoid any obstruction in expending his personal wealth, a decision that would change US national electoral campaign financing for good by removing the relative financial restraint imposed by FEC funds matching. Also in 1995, a new, fully formalized international institution, the World Trade Organization (WTO), replaced the treaty structure known as the General Agreement on Tariffs and Trade (GATT) that dated to the end of World War II—an event that might be understood as economically stabilizing, were it not for the prompt eruption of disputes between developed and developing-economy members (the as yet unresolved "Singapore issues") and the attention of anti-globalization activists, which would culminate in violent street protests at the 1999 Seattle conference.

The year 1995 was not an uneventful one politically, either. US national political volatility increased as Speaker of the US House of Representatives Newt Gingrich, capitalizing on Republican success in the 1994 midterm elections, finished crafting the insurgent conservative legislation known as the Contract with America and forced the first of a series of US federal government closures in a dispute with President Bill Clinton. The nearly two-decade-long bombing campaign of Theodore John "Ted" Kaczynski, a former University of California, Berkeley mathematician who had simultaneously renounced modern technology and taught himself to construct primitive explosives (and who had targeted academic scientists and computer stores in particular) culminated with a series of explanatory letters and the publication of the so-called "Unabomber Manifesto" by the *New York Times* and *Washington Post.* And Timothy McVeigh and Terry Nichols destroyed the Alfred P. Murrah Federal Building in Oklahoma City with a truck bomb in the most significant act of domestic terrorism in the United States then and since.

While such details merely share a broad historical context with my topic of focus in this chapter, the history of the JavaScript programming language, each of these details, from the emergence of newly privatized and newly publicly accessible Internet services, new economic governance institutions, and a new class of inexpert financial speculators, to what are still remembered today as very significant acts of domestic terrorism, involved conflicts and negotiations of technical expertise, in a broad sense, and some of them were marked by such conflicts and negotiations in the narrower sense relating specifically to computers, as well. In that sense, that broader context cannot be separated entirely from my topic here.

JavaScript as Multi-paradigm Programming Language

The Java-like language that Brendan Eich was commissioned to design for the Netscape Navigator web browser in 1995 (a task that he reportedly completed in ten days) was initially named Mocha and then LiveScript. It acquired the name JavaScript with the joint press release issued by Sun and Netscape in December of that year, which I have already mentioned. In the two decades since, Java applets have almost completely vanished from the web, and it is JavaScript that provides the main interactive element in browser pages. Sun and Netscape's joint press release reminds us just how far our current situation today is from the expectations they articulated in 1995. Though some of the rhetorical choices made in the press release are perhaps more directly reflective of competition and licensing conflicts than anything else, it is worth dwelling on just how closely the respective domains of Java and JavaScript were positioned at the time:

- "The JavaScript language complements Java, Sun's industry-leading object-oriented, cross-platform programming language."

- "JavaScript is an easy-to-use object scripting language designed for creating live online applications that link together objects and resources on both clients and servers. While Java is used by programmers to create new objects and applets, JavaScript is designed for use by HTML page authors and enterprise application developers to dynamically script the behavior of objects running on either the client or the server."

- " 'Programmers have been overwhelmingly enthusiastic about Java because it was designed from the ground up for the Internet. Java-Script is a natural fit, since it's also designed for the Internet and Unicode-based worldwide use,' said Bill Joy, co-founder and vice president of research at Sun. 'JavaScript will be the most effective method to connect HTML-based content to Java applets'."

Java would be used to create code objects including applets (that is, small applications), and JavaScript programs would connect such objects and script (that is, configure and control) their behavior, providing them with an HTML-based user interface. If this particular separation of roles (Java as application programming language vs. JavaScript as scripting language) is clear, the attention the press release also devotes to "server-side JavaScript" may cloud it somewhat:

- "With JavaScript, an HTML page might contain an intelligent form that performs loan payment or currency exchange calculations right on the client in response to user input. A multimedia weather forecast applet written in Java can be scripted by JavaScript to display appropriate images and sounds based on the current weather readings in a region. A server-side JavaScript script might pull data out of a relational database and format it in HTML on the fly. A page might contain JavaScript scripts that run on both the client and the server. On the server, the scripts might dynamically compose and format HTML content based on user preferences stored in a relational database, and on the client, the scripts would glue together an assortment of Java applets and HTML form elements into a live interactive user interface for specifying a net-wide search for information."

- "Java programs and JavaScript scripts are designed to run on both clients and servers, with JavaScript scripts used to modify the properties and behavior of Java objects, so the range of live online applications that dynamically present information to and interact with users over enterprise networks or the Internet is virtually unlimited. Netscape will support Java and JavaScript in client and server products as well as programming tools and applications to make this vision a reality."

While there is no reason that two server-side programs (or for that matter, entire code bases) cannot maintain such distinctly complementary roles as are imagined here, the question of whether JavaScript might someday be able to perform alone in both such roles seems already latent in these formulations. Indeed, there exist unambiguous records of the tension around this issue, which it does not require much imagination to find in some of the joint press release's strained locutions, which read like a parent ordering two sibling children to get along. As Eich has put it: "If I had done classes in JavaScript back in May 1995, I would have been told that it was too much like Java or that JavaScript was competing with Java [. . .] I was under marketing orders to make it look like Java but not make it too big for its britches [. . .] [JavaScript] needed to be a silly little brother language" (quoted in "Computing Conversations with Brendan Eich" 2012). Given that Java was already established, some at Netscape did not initially see any benefit in establishing and maintaining a separate language ("Computing Conversations with Brendan Eich" 2012).

Under such conditions, it is unsurprising that JavaScript was designed and implemented in haste, and with the hope that its shortcomings could be addressed in continued development. Eich brought a great deal of both organic and improvised expertise to bear on JavaScript's design. On the one hand, JavaScript's syntax is derived from the systems programming language C, by way of Java's own C-like syntax. On the other hand, Eich modeled JavaScript features like first-class functions and function closure on their equivalents in Scheme, a dialect of Lisp and so a member of a very different programming language family. Eich also adapted the prototype-based inheritance model of Self, a dialect of Smalltalk (a language and integrated programming environment designed in the 1970s for instructional and expressive computing), to provide JavaScript with object-oriented programming features. In combining these three quite different models—procedural or imperative in the case of C and Java, functional in the case of Scheme, and object-oriented in the case of Self—Eich made JavaScript a multi-paradigm language from the very start.

Although JavaScript was not the first such multi-paradigm language, such a synthesis is nontrivial in the design labor involved and in the prospects that JavaScript still presents for study even today. In synthesizing the three major programming paradigms, JavaScript certainly incorporated more complexity, even at the start, than most people would consider necessary in a language designed for novice and inexpert programmers. For novices, learning one programming model at a time (or only one model at all!) would certainly be considered more than enough. The tension between the design expertise that Eich brought to bear in creating JavaScript and its promotion as a language for inexpert users is especially interesting in Eich's own statements, which wholeheartedly endorse such promotion:

What people wanted back then (and still want) is the ability to go one step beyond HTML and add a little bit of code that makes a web page dynamic—that makes things move, respond to user input, or change color; that makes new windows pop up; or that raises a dialog box to ask a question, with an answer necessary to proceed—things that HTML cannot express. That's really where you need a programming language, but something simpler than Java or C++. Content creation should not be recondite. It should not be this bizarre arcana that only experts and gold-plated computer science gurus can do. (Andreessen 1998)

JavaScript as Translational Programming Language

Despite borrowing most of its syntax from C and Java, JavaScript can certainly be written in a way that makes it resemble Scheme (see Code Example 1) and dialects of Lisp more generally. In his foreword to the new JavaScript edition of the classic Scheme-based textbook *Structure and Interpretation of Computer Programs* (*SICP*) published in 2022, Guy L. Steele Jr. reminds readers that "JavaScript is not as distant from Lisp as you would think," and that the adaptation of *SICP* to JavaScript is not just a concession to the times (Steele Jr. 2022, xv). To its explicit synthesis of multiple programming models or paradigms (procedural or imperative, functional, and object-oriented), and to the divergent idiomaticity facilitated by the incorporation of features from very different languages, whose syntactic expressions may deform JavaScript's basically C-like syntax, we must add two other "multilingual" contexts for the development of JavaScript from 1995 to the present. The first is the development of of the ECMAScript standard, implementations of (and deviations from) the standard in major web browsers, and the ongoing, both forward and backward "translation" by which the development of the standard, its implementation in browsers, and its use in web programming are mediated. The second is the compilation of JavaScript to other, often nonrelated programming languages. I will describe each in turn.

```
function Y(le) {
    return (function (f) {
        return f(f);
    }(function (f) {
        return le(function (x) {
            return f(f)(x);
        });
    }));
}
```

CODE EXAMPLE 1: The Applicative Order Y Combinator, from Friedman and Fel-leisen (1996), implemented by Douglas Crockford in JavaScript. See Crockford (2003).

I have already mentioned the first edition of ECMA-262 (ISO/IEC 16262), Ecma International's specification for ECMAScript, which described ECMAScript as a scripting language "intended for use by both professional and non-professional programmers" ("ECMAScript: A General Purpose, Cross-Platform Programming Language . . ." 1997, 1). There exist thirteen published editions to date (ECMAScript 1, 2, 3, 5, 5.1, 6, and 7–13, excluding ECMAScript 4, which was abandoned), and the first edition's emphasis on design for nonprofessional users was retained all the way through the fifth edition published in 2009. The third edition published in 1999 included minor changes to the initial paragraphs of the section titled "Overview" (section 4), changing "A scripting language is intended for use by both professional and non-professional programmers, and therefore there may be a number of informalities built into the language" to "A scripting language is intended for use by both professional and nonprofessional programmers. To accommodate non-professional programmers, some aspects of the language may be somewhat less strict" ("ECMAScript Language Specification: Standard ECMA-262, 3rd Edition" 1999, 1). The fifth edition published in 2009 deleted the latter sentence, leaving only "A scripting language is intended for use by both professional and nonprofessional programmers" ("ECMAScript Language Specification: Standard ECMA-262, 5th Edition" 2009, 2).

The sixth edition published in 2015 made much more significant changes, which put meaningful distance between JavaScript's history and its present. To the paragraph defining a scripting language, a new sentence was prepended: "ECMAScript was originally designed to be used as a scripting language, but has become widely used as a general purpose programming language" ("ECMAScript Language Specification: Standard ECMA-262, 6th Edition" 2015). An entirely new paragraph was added elaborating this point:

> ECMAScript usage has moved beyond simple scripting and it is now used for the full spectrum of programming tasks in many different environments and scales. As the usage of ECMAScript has expanded, so has the features and facilities it provides. ECMAScript is now a fully featured general propose programming language. ("ECMAScript Language Specification: Standard ECMA-262, 6th Edition" 2015)

Since the end of the ten-year interval separating the third and fifth editions, which saw the development and then abandonment of a fourth edition, feature addition has been rapid and extensive, with the fifth edition in 2009 and the sixth in 2015 both adding significant new features. After the long, partly fallow interval from 1999 to 2009, this rapid pace of development stimulated the development of a culture of experimental implementation in which features still only in proposed or only partially and nonbindingly approved form, in published drafts and other documents relation to the ECMAScript specification, were included in beta or developer versions of major web browsers, and eventually even in some user versions. Even before reaching a developer version of a web browser like Google Chrome, such features found their way into use through the mediation of source-to-source compilers (also called transcompilers or transpilers) that rewrote JavaScript code using experimental features in a form compliant with previously published editions of the ECMAScript specification (and thus guaranteed to work in user versions of browsers). At the same time, as other browser vendors (Microsoft with its Internet Explorer browser, and to a lesser extent Apple with its Safari browser) failed to implement all the features in already published past editions of the ECMAScript specification, transpilation was used to make JavaScript code uniformly executable across browser platforms.

Source-to-source compilation is as old as the history of higher-level programming languages themselves, with examples dating all the way back to the 1950s. The especially vigorous, even frenetic pace of such activity in web development and other JavaScript programming today merely hyperanimates the long history of *translation metaphors* through which the history of digital computing itself can be traced.[14] Yet Java-

Script programming may well be unique and unprecedented in the range and scale of such activity, if not in its mere fact. It is appropriate indeed that the most widely used source-to-source JavaScript compiler used to rewrite JavaScript code to conform to different ECMAScript specifications has the name Babel ("Babel: A Compiler for Writing Next Generation JavaScript" 2016).

We have not mentioned the many other programming languages that provide the option to transpile to JavaScript in addition to their original targets. An authoritative list includes not only many variants best described as JavaScript subsets, supersets, or extensions (many of them with names relating to coffee, such as the large CoffeeScript family), but compilers that will take in code in C/C++, Java, Perl, Python, Ruby, C#, Scala, Clojure, OCaml, Haskell, and other both major and minor, older and newer languages and rewrite it in JavaScript (Ashkenas 2016). Here too, it is unlikely that anything else of this range and scale has ever been seen in the history of software programming. In a domain that is and has always been defined by constant translation, JavaScript programming culture can be distinguished as exceptionally *translational*.

Node.js

A major, perhaps *the* major factor in the rapid expansion of JavaScript's domain is the Node.js project, which dates to 2009. Like its successor, the Deno project that emerged a decade later,[15] Node.js is a JavaScript runtime environment disembedded from the browser software application and written much as scripting languages like Python and Ruby are written: a developer may use command-line utilities, including a REPL (Read-Evaluate-Print-Loop) and debugger, along with a software application text editor or Integrated Development Environment (IDE) to write and test programs locally on their own computer, but outside the browser application environment. The Node.js interpreter can run on a server, as well as on a desktop PC and in other software and networking contexts, and it can be embedded in a wide range of computing devices. While the earlier popularization of so-called Ajax (Asynchronous JavaScript and XML) programming techniques, closely associated with Google's Maps products, certainly got the ball rolling in this respect, it may be Node.js more than any other single factor that has transformed JavaScript from a scripting language initially used disproportionately to create "annoyances" like browser pop-up windows ("Computing Conversations with Brendan Eich" 2012) into something approaching a full-fledged systems programming language. Arguably, the Node.js project has revived, reactualized, and then realized the promise of the forgotten chapter of JavaScript's history marked by Netscape's early imagination and exploration of server-side JavaScript applications. For the first time since the appearance of the World Wide Web and the software browser application, Node.js unifies so-called front-end and back-end web development, so that so-called full stack developers, who write code for both user-facing and data-processing components of an application, can use a single programming language for both tasks.

The applications of Node.js go beyond the server (though in that context, we should also mention the displacement of XML, as used in the earliest Ajax techniques, by the JavaScript-associated JSON [JavaScript Object Notation] data exchange format). Web application frameworks written in Node (Express.js and Meteor are two of the best known early examples) rapidly eroded the popularity of Rails, the Ruby-based application framework that dominated web development from the late 2000s onward. The Node-based Electron framework is now widely used to develop platform-agnostic desktop GUI applications (that is, conventional applications that a user downloads and runs on their own machine, rather than using in a browser window). Node's popularity also explains Apple's inclusion in 2015 of JavaScript as one of the operating system languages of macOS (formerly OS X), usable for interapplication communication, as well as the inclusion of JavaScript "bridges" in the Apple iOS and Google Android operating systems for mobile platforms, enabling compilation to a JavaScript bound to the native system programming languages of those platforms (C, C++, Objective C/Swift, and Java).

The Frameworks

For the reasons mentioned above, it is Node.js, more than anything else, that has driven the recent hyper-professionalization of JavaScript programming, removing the language quite decisively from that portion of its design origins that emphasized accessibility to inexpert and nonprofessional programmers. A development separate from the Node.js project, but of nearly equal impact on both JavaScript's expansion and its hyper-professionalization, is the proliferation of other JavaScript-based web application frameworks *not* directly designed for or implemented in Node. The Ember.js, AngularJS, Vue.js, and React frameworks have provided web developers with very sophisticated, unified abstractions of the three core browser technologies (HTML, CSS, and client-side Java-Script itself) that have made possible great leaps in both the sophistication of web applications and the creativity and productivity of those who write them. But they have also quite decisively propelled web development beyond the domain of accessibility imagined for the Web when it first appeared, which may have persisted in real terms for as long as ten years after 1995, in the sense that wage-earning web programming techniques could still be learned through casual and part-time training or retraining.

The professional construction and maintenance of web sites today requires both initial and ongoing training, and requires a level of skill maintenance and retraining, that puts it well out of reach for virtually anyone who is unable to devote her- or himself to its full-time pursuit—even academic researchers, excepting those who study and teach web technologies as a well-defined and well-developed technical research specialty. It is not the mere passage of time that makes the humanities-based wave of emancipatory hypertext theory of the 1990s, for example, seem so quaint,[16] and makes its reinstantiation in a later "digital humanities" movement seem so disingenuous or so guileless, depending on whom one asks. It is not that JavaScript no longer stands for technical improvisation, but that in what I have been calling improvised expertise, expertise is now both unambiguously and unambivalently the agent of improvisation, instead of its object.

JavaScript Fatigue—and Other Futures

This shift has not been universally (or even broadly) welcomed. Indeed, the frenetic pace of change, if not the levels of skill required, is clearly a burden even to well-trained and experienced full-time professional developers. In early 2016, the phrase "JavaScript fatigue" began appearing in social media posts, blog writing, podcast discussion, and discussion at professional JavaScript developer conferences and briefly dominated such discussions as a collective preoccupation.[17] The developer who began circulating the phrase had lamented the "confusing nest of build tools, boilerplate, linters, & [other] time-sinks" that the individual and combined profusion of new language features, transpilers, frameworks, and other developer "tooling" (that is, custom applications for various programming tasks) represented: an entire preliminary phase of assessment and labor that was required before a JavaScript web application could even begin to be designed (Clemmons 2015).

It is this profusion, which many JavaScript developers find suffocating, that my intentionally lighthearted title "JavaScript Affogato" is intended to name, referring to the Italian dessert consisting of ice cream or other sweets "drowned" in espresso coffee.[18] A soberer assessment would be bound to remind us that at some level, howsoever mediated, this hypertrophy of labor, and all the structural and personal strains that go with it, reflect or perhaps co-constitute the extensive economic violence of the interval beginning in 2008. Professional programmers have been among the few labor-market beneficiaries of an extended era of austerity and generalized economic pain and suffering, and such luck is nothing if not equally a curse. Certainly the significant priority placed on JavaScript, in particular, and its concomitant growth during this period, reflects the priorities declared by investment patterns focused on the user-facing "app" as a cultural token, associated with making and building things as "free" labor (free as in freedom) and adaptation to austerity, and as a new object of financial engineering.

In 2014 Brendan Eich's resignation as CEO of the Mozilla Corporation, only nine days after taking the position, was described by a writer for the *New Yorker* as "the least surprising C.E.O. departure ever," given that Silicon Valley was "a region of the business world where social liberalism is close to a universal ideology" (Surowiecki 2014). (Eich's having donated to an anti–marriage equality campaign supporting California's ballot Proposition 8 in 2008 was a fact known beforehand, which became newly controversial upon Eich's appointment.) It might be more accurate to say that in a broader context, the episode reflects the fundamental confusion of the specifically cyberlibertarian politics of Silicon Valley investment and management culture, which borrows ideas freely, but mostly unreflectively and unsynthetically from both statist left-wing and antistatist right-wing political platforms, in ways that seem to reflect the startup culture's ambivalence about its own expertise. Though a great deal of work remains to be done to articulate the meaning such contexts lend to topics such as my own in this book, such social dynamics cannot be delinked from the technical history of the artifacts designed and produced in contexts determined by them—even, or perhaps especially, such an artifact as a programming language.

———

In this chapter, we have performed a second close study of a specific programming language, the second part of a dual case study whose contrasts illuminate the complexity of the links among software, automation, and expertise. From this vantage point, looking back, we see that the improvised expertise attending the historical development of JavaScript and its usage cultures had an ancestor in the improvised *in*expertise, or amateurism or generalism, of the historical development of the Snobol languages and their own usage cultures. Returning to a broader context, in the final chapter of this book, which serves as a coda and afterword, we gather this book's discussions of software project management commentary into a concluding analysis of contemporary literary articulations of the currently dominant form of management-oriented software automation.

SIX

DevOps Fiction

WORKFORCE RELATIONS IN
TECHNOLOGY INDUSTRY NOVELS

Describing the typical evolution of a software project management disaster, Bill Palmer, the narrator of Gene Kim, Kevin Behr, and George Spafford's *The Phoenix Project,* laments that "because no one is willing to slip the deployment date, everyone after Development has to take outrageous and unacceptable shortcuts to hit the date. [. . .] [I]t's always IT Operations who still has to stay up all night, rebooting servers hourly to compensate for crappy code" (Kim, Behr, and Spafford 2014, 53). Anchored to the middle of Facebook's golden years, a moment when the software developer was still imagined an economic hero,[1] Palmer's allegorization of the conflicts between the makers and the maintainers of software infrastructure is enthusiastically one-sided, unhesitating in its indictment of "idiotic" Development as the source of the industry's problems.

That software developers, as one professional class, are locked in conflict with IT professionals, as another, equally distinct group, may be surprising to those with little knowledge of the software industry and its ways. That this conflict would be memorialized in a sub-area of contemporary US literature, of all places, might well come as a second surprise. Yet at the present historical moment, we might want to programmatically set aside such surprise in the interest of a truly collective education. At this point, our situation is what it is. Andrew Smith is right to argue that "by accident more than design, coders now comprise a Fifth Estate and as 21st-century citizens we need to be able to interrogate them as deeply as we interrogate politicians, marketers, the players of Wall Street and the media" (Smith 2018, n.p.). Smith's observation is a recent one, prompted by the rise of the new platform monopolies (chiefly, if not exclusively Facebook) and the augmentative and consolidating reorganization of the slightly older ones (Google, Amazon.com) during the 2000s, but this should not be a new concern. As Robert Britcher put it more than two decades ago, "Behind the scenes, software runs everything from banking to space exploration. But the number of people who know how software is written is small. Most of us are not talking, or we are saying 'Trust us'" (Britcher 1999, 17). If we collectively—that is, the truly vast majority who do not create software for a living—hope to understand in its fullest dimensions the role that the work culture of Silicon Valley has lately come to play in our general culture, then among other things, including the fundamentals of computing and the procedures of software development, we need to grasp the fundamentals of the management modes and theories by which software developers are managed, both by those who supervise them and by those who work alongside them. Without such an understanding, I argue, we cannot make headway in collectively managing the software industry, something the trajectory of recent years has made it urgent to do.[2]

This final chapter surveys the literary expression of the spontaneous management theory associated with the concept of "DevOps," or "development operations," which emerged during the heady 2000s, a decade in which Silicon Valley industry crashed, returned to life, and was boosted, rather than constrained, by the Great Recession of 2008. It proposes that we regard DevOps fiction as a legible category of contemporary literature, meaningfully distinct from Silicon Valley fiction as its superset. Rather than sensationalizing, for better or for worse, either the elite consumption of new digital, social, or platform products or their production by elite software engineers, DevOps fiction imagines the management of routine maintenance without which neither is possible. As such, it enacts the shift of focus from innovation to sustention or sustainment that Andrew L. Russell and Lee Vinson have argued would be both judicious and necessary.[3]

Indeed, the Silicon Valley novel itself is already deep into its second wave, though the chronicles of the 2000s boom to be found in Anna Yen's *Sophia of Silicon Valley* (2018), Doree Shafrir's *Startup* (2017), Rob Reid's *After On* (2017), Ann Bridges's *Rare Mettle* (2016) and *Private Offerings* (2015), Dave Eggers's *The Circle* (2013), Michael S. Malone's *Learning Curve* (2013), Keith Raffel's *Dot Dead* (2013), Robin Sloan's *Mr. Penumbra's 24-Hour Bookstore* (2012), and Greg Bardsley's *Cash Out* (2012) can be difficult to distinguish from fiction produced during the run-up to the dot-com crash, such as Ellen Ullman's *The Bug* (2003), Po Bronson's *The First $20 Million Is Always the Hardest* (1997), Pat Dillon's *The Last Best Thing* (1997), and Douglas Coupland's *Microserfs* (1995). Martin Paul Eve and Joe Street describe the form as "a sub-genre of a type of platform-capitalism fiction" that takes its object—"Silicon Valley" as a historical moment, more than just a place—"as a parameter of the present that signifies an unfolding failure of modernity" (Eve and Street 2018, 83). For the most part, their explication of this thesis sticks closest to literary historiography (the interplay and trade of modernism with postmodernism in literature), ranging to the historic enfolding of Silicon Valley industry within US defense imperatives where the bodies of work of specific writers reflect on that entanglement; this is consistent with the focus on Silicon Valley industry products and effects, in so much of this work.

DevOps fiction is, then, a subgenre of the subgenre of Silicon Valley fiction, whose object is not the drama of late capitalist modernity (and the never-ending question of what's next) as much as its stasis, and the micrological struggle of that stasis with the charisma of innovation. In its focus on the maintenance of the production of software and the internal "ops" of software production, DevOps fiction is perhaps best understood as a literary branch of the line of project management commentary that begins with Gerald M. Weinberg's *The Psychology of Computer Programming* (1971) and Fredrick P. Brooks Jr.'s *The Mythical Man-Month: Essays on Software Engineering* (1975) and includes such recent reflections as Robert N. Britcher's *The Limits of Software: People, Projects, and Perspectives* (1999) and Scott Rosenberg's *Dreaming in Code: Two Dozen Programmers, Three Years, 4,732 Bugs, and One Quest for Transcendent Software* (2007). While Brooks's *The Mythical Man-Month* is widely and justly considered a classic, it is Weinberg's *The Psychology of Computer Programming,* the purview of which is far broader than its title suggests, that opened the door we walk through here, less because it was published first than because it was originally conceived as a novel (Weinberg 1998, xi).

Britcher's *The Limits of Software* is Weinberg's successor in this respect. An intricately structured and self-consciously literary, if also casual and mostly expository fiction blending the expansive introspection of the Montaignian essay with the constrained worldliness of the New Journalism, it presents itself as "an entertainment aimed at the truth" in the domain of software, where the personal and technical "are bound more tightly than in any other form of technology" (Britcher 1999, xvi). Interleaved with the fictionalized journal entries of a shadow narrator and alter ego of the book's author (a systems designer and project manager retired from IBM), its seventeen episodic chapters are divided into two parts in counterpoint, setting the infrastructural emergence of software engineering in the 1960s and 1970s against its financialized glamorization in the 1990s, a "new era" defined by "software as the stuff of high expectations and money." This apparently "happy marriage," we learn to no particular surprise, "has led to one disappointment after another" (Britcher 1999, xvi), culminating in the catastrophe of the US Federal Aviation Administration's Advanced Automation System (AAS), a software project that began in 1981 and collapsed thirteen years later in "the greatest debacle in the history of organized work" (Britcher 1999, 163). More concretely and more insistently than anything else in the admittedly underpopulated domain of software engineering introspection, Britcher's suite of essays describes software as recursive automation, or the automation of automation, in that newer software absorbs and incompletely digests old software, rather than replacing the latter, in a process whose complexity can grow exponentially, proving itself ungovernable time and again. "We may never recover," observes the shadow narrator (whose journal entries are printed in white text on a black background) "from the beginnings of digital computing and programming [. . .] We cannot afford to pull the rug out from under the automation created just one or two or three decades ago. No institution [. . .] can accord to *replace* its automation" (Britcher 1999, 190). The FAA's AAS failed "largely because of our thirst for automation and a belief that the easier life is a better one," and while objections to this canalization "are coming along," it "is late in the game. Officially, we are in love with this latest form of automation" (Britcher 1999, xvi).

Like any other love affair, it is a world of pain, and save for the odd detail here and there dating *The Limits of Software* to the beginning of the end of the dot-com boom, the struggle it depicts is fully contemporary more than two decades after its publication. Each of the big tech companies has its own name for the commitment to keeping their vast service infrastructure upright and available. Google calls it Site Reliability Engineering, whose mission Cade Metz, writing for *Wired*, described as "Don't get IT people who specialize in running Internet services to run your Internet services" (Metz 2016, n.p.). While the association of IT system administration with rote work, grunt work, and boring work is a liability, in that most younger software engineers and many older ones vastly prefer building flashy new products for external use, it is an unavoidable liability, and to address that liability effectively means making routine operations work itself more interesting. By deliberately dividing a site reliability engineer's time between making and maintaining, with an incentivizingly hard limit on the proportion spent on the latter, a company partly merges two personnel domains whose conflicting priorities create a core management problem. For a company to treat IT operations as software development, rather than infrastructure maintenance, is to put its higher-status or "sexier" craftspeople on the knowledge work equivalent of an assembly line (Metz 2011, n.p.).

Advances in cloud computing infrastructure during the 2000s, from which emerged Amazon Web Services, Microsoft Azure, and offerings by Google, IBM, and Oracle, among others, made possible such new vendor models as software as a service (SaaS), along with broader ("platform" or "back-end" or simply "infrastructure" as a service) and narrower variants (individual functions as a service, in the "serverless" computing provided by AWS Lambda or Azure Functions). Traditional IT systems administrators or "sysadmins," trained to administer private organizational networks running on locally provisioned server hardware, had to adapt to provisioning virtual servers through a management interface layering new abstractions on top of hardware now well beyond reach. In some ways, the traditional sysadmin was reduced to a user—different in kind, to be sure, from the nontechnical professionals and other knowledge workers on behalf of which the sysadmin had always managed resources, but no less meaningfully deskilled by the emergence of new modes of virtualization and automation. There is no way around the fact that developers took part of the sysadmin's job, not necessarily willingly, and while that involved some additional training, the burden of adaptation was borne disproportionately by the sysadmin.

In addition to this reorganization of work, DevOps is associated with a variety of other recent transformations in the production of software application and services. At all of the largest (and many smaller) companies, the traditional software application release cycle, for many decades defined by alpha and beta testing phases prior to releases at a pace of one major version each year or interval of years, has been displaced by the continuous delivery of production changes many times each day (at a rate measurable in the thousands or even tens of thousands), with new micro-versions injected into use live, without meaningful service interruptions. In 2001, experiments with lightweight incremental and iterative software development methods were unified by the publication of a "Manifesto for Agile Software Development," which in addition to the continuous delivery of software included among its twelve principles adaptability to change in customer requirements, minimizing wasted labor and time, the self-organization and autonomy of development teams, and emphasizing trust over supervision in managing developers. The roots of many of the latter ideas are in the Toyota Production System developed beginning in the late 1940s and its just-in-time or lean production principles, emphasizing low inventory, minimization of waste, and the limitation of work in process, which is "pulled" rather than "pushed," to prevent it from piling up. Indeed, the importance in Agile practice of concepts like kaizen (ongoing incremental improvement) and organizational systems like kanban (signboard) suggest a meaningful management-theoretical, if possibly counterintuitive continuity between the early postwar history of auto manufacturing and late twentieth- and early twenty-first-century software production, as critical and integral economic phenomena and economic-historical domains of substantial symbolic value. While terminologically speaking, the segmentation and then reunification of Dev and Ops is a software industry process, the flow of work from "upstream" Development to "downstream" Operations mirrors older paradigms for the production of machines as material goods.

The term "DevOps" emerged in June 2009, when John Allspaw, then leading the Operations Engineering group of the image hosting service Flickr.com, and his colleague Paul Hammond gave a presentation titled "10+ Deploys Per Day: Dev and Ops Cooperation at Flickr" at the Velocity

Web Performance and Operations Conference in San Jose, California.[4] Arguing that the growth of web infrastructure in many cases obscured the traditionally observed difference between fundamental or back-end computing system infrastructure and the user-facing software applications running atop it, Allspaw and Hammond described Flickr's experiments with controlling a culture of "finger-pointyness" grounded in the mutually exclusive, hardware- and software-based responsibilities of sysadmins who blamed production problems on developers ("It's not my machines, it's your code!") and developers who blamed them on sysadmins ("It's not my code, it's your machines!").[5] As the title of the presentation, the phrase "10+ deploys per day" functions as both a challenge and a counterintuition. Traditionally, a software deployment, or the release, installation, and activation of a new version of a software application or service, was an intricate and time-consuming operation. Until the 1990s, it might be undertaken only once a year or less. During the interval set aside for deployment, new development might stop altogether, as developers who had written the code for new features and revised code to eliminate bugs went home, or even took their vacations, while the sysadmins took over. In the 1990s, a newly public internet and the browser-based web infrastructure sitting atop it made it possible to provide publicly accessible nightly builds, or automated deployments of an application or service incorporating all changes made during a day of work. While there remained a meaningfully separate concept of a release build thoroughly tested and accompanied by announcements, change lists, and other public documents, the availability of infrastructure for providing nightly builds changed the tempo of software development work quite radically. And since automation in computing might be termed recursive, in that each new compression of the time and space of production takes as its object not a manual precedent but a previous achievement of automation, once software could effectively be released every day, it was only a matter of time before the workday itself was eclipsed as a segmenting measure.

In "10+ Deploys Per Day: Dev and Ops Cooperation at Flickr," Allspaw and Hammond began with a typical Silicon Valley business canard: "the business requires change"—less memorable, perhaps, than "move fast and break things," but no less directly derived from the Californian ideological inflection of Schumpeter's gale. To that they added immediately, however, that change "is the root cost of most [service] outages" on a platform. The solution, they proposed, was neither to discourage change (the traditional role of Ops) nor to allow change to proceed uncontrolled, but to build new development tools and a culture of their use to "allow change to happen as often as it needs to." The emergence of DevOps thus also marks the expansion of that portion of the technology industry devoted entirely to providing applications and services to software developers and their sysadmin colleagues themselves, rather than to any end user. (Sydney-based Atlassian, the success of which produced Australia's first tech billionaires at its IPO in 2015, is a good example: in February 2019, Nellie Bowles of the *New York Times* described it as "a very boring software company [. . .] [that] develops products for software engineers and project managers" (Bowles 2019).)

Nick Srnicek has described the post-2008 tech boom as a project of "surplus capital seeking higher rates of return in a low interest rate environment" (Srnicek 2017, 86). Describing platforms as a "new type of firm [. . .] providing the infrastructure to intermediate between different user groups" in this specific context, Srnicek emphasizes that they tend to emerge from "internal needs to handle data" inside a company, subsequently scaled up into consumer products and services (Srnicek 2017, 43, 48). While the reconciliation-in-separation of Development and Operations is theoretically compatible with the work arrangements and management approaches characteristic of any of the five types of platforms Srnicek identifies, including advertising platforms (Google, Facebook), cloud platforms (Amazon Web Services), lean platforms (Uber, Airbnb), product platforms (Spotify, Zipcar), and industrial platforms (General Electric's Predix, Siemens's MindSphere), it is most directly linked to the emergence of cloud platforms, among other things as a deskilling extrusion, distribution, and encapsulation of the technical labor of the sysadmin. Certainly it is best located within the constrained category of the "tech sector," a term Srnicek deliberately subordinates to the broader scope of a "digital economy"; nevertheless, Srnicek's emphasis on competition among companies as the agents of capital, and his deprivileging of the agency of labor on account of its profound weakness, in this period, furnishes us with relevant parameters of analysis. DevOps is not the union of Development and Operations as much as the management of the former for the benefit of the latter, and to some extent *by* the latter. Like so much in an age of what Srnicek calls "a hegemonic model" of disruption (Srnicek 2017, 5), the dynamic of this process is more colonization than consilience.

Like Srnicek's *Platform Capitalism,* Dan Schiller's *Digital Depression: Information Technology and Economic Crisis* (2014) took as a point of departure the contemporary moment, historicizing events leading to it and potentially leading away from it into the future, though the scope of Schiller's book (and the broader research endeavor of which it is a part) is far broader. "Today," Schiller began, "a technological revolution is wrapped up inside an economic collapse: Whence came this digital depression? The same edition of the *Financial Times* that headlined 'deflation fears' in November 2013 also carried a front-page story heralding the good fortune of Twitter's founders, when they became overnight billionaires following the company's initial public offering. What accounts for the emergence of this pole of relative growth, which is so evident, even amid the slump?" (Schiller 2014, 5–6). Schiller described the 2000s as marked by a global crisis "paradoxically" originating in the United States as the site of regenerative innovation in advanced information and communications technology (ICT); by a contradiction between public discourse and economic realities, as what Schiller termed "ICTs" were transformed from "a source of general economic uplift" into a driver of a Great Recession; and by continuation of the crisis, despite proclamations of a recovery beginning in 2009 (Schiller 2014, 1–2). Schiller's position, heterodox in relation to liberal and radical economic theory (for example, the work of David Harvey and Vijay Prashad) as much as to mainstream economics, is that ICTs are central to the historical development of capitalism from the mid-twentieth century onward, and that in the course of what Brenner (2006) has called "the long downturn," the economic role of ICTs was a restructuring role, as capital used ICTs successively to restructure production, then finance, then military expenditure, then finally the ICT sector itself.

This final transformation (final for now) is as significant as what preceded it. "What we call 'the internet'," Schiller argues, "did not storm the world fully fledged": the construction and maintenance of internal corporate computer networks was already a growth industry, by the time the World Wide Web appeared in the mid-1990s, and the effect of the latter was less to introduce than to both intensify and routinize a violent "rerouting" of "existing commodity chains." "Web services and applications were able to hopscotch across a growing installed base of desktop personal computers, to and from what, before this, had been separate, specialized networks operated by large organizations [. . .] This unexpected bridging introduced potent and [. . .] intrusive network effects. Existing nodes of market power were disrupted and had to be recreated in different form" (Schiller 2014, 80–81). To judge by the attestations of those who found their technical training being routinized, redirected, or replaced during this transition, the effects were swift and often savage. Roger Clarke argues that system administration as a profession "peaked 1980 to 2005, and has been in decline as outsourcing of enterprise platforms took hold, and organisations ceded control of their information infrastructure and put themselves at the mercy of service-providers" (Clarke 2012). In the testimonies assembled by Clarke, the latter included Microsoft first of all, but providers of remote cloud, hosted, and co-located data center-based services more generally, the transfer of attention to which sysadmins describe explicitly as deskilling: "Skills I studied and trained for are all of a sudden redundant [. . .] deskilling increases as more tasks are taken away and 'standardised' [. . .] I have no control over my career anymore. I thought I could do this for the rest of my working life, but I think the role of the sysadmin is not as respected or as relied upon as it once was. Like the old lawn mower that used to do a great job as it was pushed around, we are just old smelly dinosaurs in the garage of modern IT."

Britcher's description of software as the automation of automation is equally pertinent here, and this conceptualization carries and extends an intellectual tradition beginning with nineteenth-century debates in political economy triggered by the thoroughgoing mechanization of manufacturing, as Amy Sue Bix summarizes it in her study of "technological employment talk in the United States" (Bix 2002, 8). Derived from Jean-Baptiste Say's *Treatise on Political Economy* (1803), "Say's Law" or the "Law of Markets," when applied to the mechanization of manufacturing, stipulated that in the long run, mechanization augmented employment and otherwise proved a net benefit. Notable responses to Say's formulations included Jean Charles Léonard de Sismondi's suggestion that "mechanization would displace workers faster than consumption could grow" (Bix 2002, 27) and David Ricardo's 1823 revision of *The Principles of Political Economy and Taxation,* which qualified or retracted his reinforcement of Say's views in the book's first edition (Bix 2002, 28).

In the twentieth century, Bix dwelled on *Displacement of Men by Machines: Effects of Technological Change in Commercial Printing* (1933) by the Columbia University economist Elizabeth Baker (1885–1973), who used the term "technocultural employment" rather than "technological unemployment" to mark the cultural dimension of adoption of disruptive technologies (Bix 2002, 33). Also noteworthy was the report "Automation and Technological Change: Report of the Joint Committee on the Economic Report to the Congress of the United States," published in 1956 and derived from congressional hearings the previous year. Bix described these hearings as focused "on the economical and social effects of automation, with special attention to 'possible and probable displacement of personnel'" (Bix 2002, 243), noting that "the committee asked speakers to clarify the strange word *automation* (not yet included in major dictionaries). In fact, experts had not yet even agreed on terminology, speaking variously of 'atomizing' or 'automatizing' as well as 'automating'" (Bix 2002, 244). Despite the novelty marked by this lexical exploration or fumbling, the committee's charge was tempered, at least rhetorically, by the vicissitudes of public mood and perception accompanying economic phenomena in US culture. Bix contrasted the 1955 hearings with the 1939–40 Temporary National Economic Committee (TNEC) study of market concentration and monopoly power, which had included a focus on the role of technology, noting that while in 1939, "record job loss had been fresh in people's minds," by the early 1950s the United States was approaching full employment, at least by the usual measures (Bix 2002, 249). "Although the TNEC members had interpreted labor's prospects in the most pessimistic tones," Bix observed, "their postwar counterparts ultimately concluded that, given current fiscal well-being, automation posed no real danger" (Bix 2002, 249).

A para-institutional model of analysis was represented by the Ad Hoc Committee on the Triple Revolution, an association of academic experts and activists that articulated a left-leaning political analysis of the relationships among cybernetic automation, nuclear weaponry, and human and civil rights, and made what Bix called an "attempt to create a coherent approach for dealing with technological unemployment" (Bix 2002, 269). Bix was acerbic in her characterization of both this endeavor's assumptions and its prospects. "Far from advocating rebellion," as Bix put it, "the Ad Hoc Committee urged Americans to come to terms with automation as destiny [. . .] Rather than depending on employment for survival, every family should naturally claim the 'right' to a guaranteed income. The overly socialist implications of such radical proposals ensured their rejection" (Bix 2002, 269). (In *Dismantlings: Words against Machines in the American Long Seventies,* Tierney [2019] provides a useful counter to Bix's polite but curt dismissal, observing that despite its optimism about the prospect of democratic social change through automation, the Ad Hoc Committee soberly noted that "large-scale social betterment would be a difficult long-term project" [Tierney 2019, 85] and that it articulated the "forking path of technological ethics" at its own moment, the mid-1960s, while proposing a future disentanglement of cybernetic automation from racism and militarism as well as straightforward economic exploitation.)

More recently, Benanav (2020) has argued that a new "automation discourse" emerging in the 2010s responds to a genuine economic phenomenon, chronic underemployment, but provides it with the specious and "simply false" explanation that "runaway technological change is destroying jobs" (Benanav 2020, x). In correcting this error, Benanav makes four counter-arguments. First, that the "long downturn" in the world economy since the 1970s was not produced by outsized leaps in technological progress. Rather, it is the product of measured and steady "ongoing technical change in an environment of deepening economic stagnation" amid persistent industrial overcapacity (Benanav 2020, 11). Second, that the effect of such change has been persistent underemployment, or "chronic labor underdemand," rather than spectacular mass unemployment imagined as a consequence of technological entrechats. Third, that even if "full" automation of production could be achieved, that would not "automatically entail the adoption of technocratic solutions like universal basic income" (Benanav 2020, 12). Fourth, following the arguments of "the original theorists of post-scarcity," from More and Étienne Cabet to Marx and Kropotkin, Benanav argues that realizing a post-scarcity society will require social struggle, but not the full automation of production (Benanav 2020, 83).

Though Benanav's book is deeply critical of the "automation theorists," deploying its critique in the witheringly acerbic prose style familiar to any reader of the journal *New Left Review,* Benanav pointedly notes that he is "more sympathetic to the left wing of the automation discourse," which he calls "our late-capitalist utopians," than he is to their critics elsewhere on the political spectrum (Benanav 2020, 11). In my view, this preserves something else valuable in the work of the "automation theorists," along with their principled political belief in the possibility of change. That is their recognition of the technical feasibility of full or complete automation as intrinsic to the technical logic of software, as separate from its feasibility or completability in either implementation or application, and despite their overestimation of real capacities for truly working, effective, and useful automation.

Yet Benanav's book makes little mention of computing, software, or programming in their specificities, notwithstanding an equally pointed mention, in its preface, of having interned at a series of 1990s startup companies "writing HTML and JavaScript [*sic*]" (Benanav 2020, xii). Benanav's intervention very clearly rejects the sudden-death drama of the imagined transformations wrought by software—which I can only agree with, because that transformation, as I myself have argued in this book, has been steady but slow, gradual, and lengthy, accruing over what is now eight decades. In the end, Benanav's book does bracket out the specific contributions of computerization or softwarization to the long downturn, something that doesn't fully make sense except as a possibly necessary rhetorical consequence of its critique of the automation theorists and their technocratic solutions. For that reason, in my view, Benanav's arguments need supplementation by the work of someone like Schiller, who argued that other, similar accounts, even those by liberal and radical economists and political thinkers (Schiller has singled out David Harvey and Vijay Prashad), have intentionally or unintentionally minimized the economic role of what Schiller calls information and communications technologies (ICT) in the growth of manufacturing overcapacity and the secular stagnation that followed.

In a subsequent intervention, Munn (2022) distances himself not only from the entire twentieth-century span of what he calls "technopessimism," from Keynes to Louis Stark, Kurt Vonnegut, and Marc Pelegrin, but also from the most recent of the "automation optimists" in a line of left-leaning and Marxist political thought, including Aaron Bastani, John Danaher, and the co-authors Nick Srnicek and Alex Williams (effectively, the critiques of Benanav and Munn are aimed at the same targets).[6] "While the automation optimists and the technopessimists may differ on the implications of this shift," Munn (2022) argues, "they share the same underlying assumption of technical takeover. Whether embraced as a bright future or rejected as a dark dystopia, the prognosis is remarkably consistent: full automation will fully replace the human" (Munn 2022, 15).

Like Benanav's arguments, Munn's arguments are entirely reasonable and I have no reason to dispute them. Munn does not, of course,
deny that automation technologies actually exist and are actively in
use. Rather, Munn argues, first, that their implementations and applications are always incomplete, "piecemeal rather than total" (Munn 2022,
2). Second, Munn argues that automation technologies are unevenly
distributed, operating differently in different contexts ("cultures and
locations") and impacting "particular races and genders rather than a
generic humanity" (Munn 2022, 2–3). Again, these arguments, rejecting
what Munn calls "the myth of autonomy," "the myth of automation everywhere," and "the myth of automating everyone" respectively, are entirely reasonable. I do not dispute them here. I agree with Munn when he
states that "full automation [. . .] never arrived. Total autonomy was never
achieved. Instead we are left with a future of work that is half-baked, an
array of semiautomated systems and processes that need augmentation
and intervention" (Munn 2022, 44).

However, I'd suggest that the ferocity with which Munn debunks the
term and concept "full automation" leads him to neglect the technical
reality of the technical logic of software as the automation of automation, and I would characterize this as a Simondonism latent in media
studies approaches to this topic, meaning a well-intended misreading of
that technical reality and a distracted caviling over the mostly irrelevant
question of its perfectibility. (By "Simondonism" I mean the philosophical skepticism expressed in the influential work of the French philosopher Gilbert Simondon, mentioned in the introduction and in chapter 1,
whose 1958 study *Du mode d'existence des objets techniques* [*On the Mode
of Existence of Technical Objects*] rejected automation as a "myth.") In
this book, I have adopted and adapted the phrase "recursive automation"
as an alternative to the concept Munn rejects. As a concept, recursive
automation may be technically less than one hundred percent precise, if
only because professional mathematicians and computer scientists can
argue, or else cavil, for a long time about a precise definition. As I have
acknowledged, mine is a generalizing and generally looser usage of the
term "recursion" than a strictly technical perspective might permit. But
that is deliberate, not accidental. As a phrase, "recursive automation" is
effective in capturing the specifically layered and intensifying dynamics
of software automation, which is an automation of automation in all but
the very last instance (the hardware layer)—something that Munn fails

to register in an otherwise valuable and effective study and a persuasive rhetorical performance. Another advantage of this generalization of the concept of recursion is that it marks a limit to what a functioning society can bear, in the automation of automation: that we are quite capable of "blowing the stack," as programmers call it, meaning exhausting the resources of the system, with potentially catastrophic consequences. As a legibilization of the technical logic of software, recursive automation is the counterpart of Munn's "spotty automation" and "shitty automation" (Munn 2022, 33, the latter phrase borrowed from the journalist Brian Merchant), terms that accurately mark the fragmented and partial character of actually existing automation *in its applications,* rather than in the potential contained in the intrinsic logical mechanics of software. I find this a necessary supplement to the type of work Munn's book represents.

Munn's rhetorical investments also lead him into some corners. "Unseen and unacknowledged in the spectacle" of the February 1946 public demonstration of the ENIAC computer at the Moore School of Electronics at the University of Pennsylvania, Munn reminds his readers in a chapter titled "Automation's Gendered Inequality," "was the gritty and groundbreaking labor of the women who had programmed it. The six women operators 'who had crawled through the machine's wires and vacuum tubes to enable so-called acts of machine intelligence' were never mentioned."[7]

In manipulating this anecdote, quoted from another source, Munn appears to suggest that a longstanding erasure of the role played by Jean Bartik, Kathleen Antonelli, Marlyn Meltzer, Betty Holberton, Frances Spence, and Ruth Teitelbaum in programming the ENIAC computer, and the lesson of the feminist historiography of computing that has retrieved this history, means that as automation, the programming performed by the so-called "ENIAC women" is somehow of a different order or quality than programming as automation in the usual sense. It is not. Programming is automation when it is performed by the nearly exclusively white "computer boys" who took over after the 1940s, in the midcentury masculinization of programming that Nathan Ensmenger has chronicled. Programming is no less automation when it was, and is, performed by a programmer whose gender is not male or whose racial identity is not white. Certainly, the forgetting of Bartik's, Antonelli's, Meltzer's, Holberton's, Spence's, and Teitelbaum's contributions to the history of computing is a historiographic injustice. But if the production of specific modes of automation is the production of specific modes of economic or political-economic injustice, as Munn (2022) also suggests in his book (despite rejecting the possibility of their totalization), then the participation of programmers of any gender and color in that process cannot be less unjust than the participation of the white men who converted programming to a birthright privilege in the 1950s. As the first programmers, Bartik, Antonelli, Meltzer, Holberton, Spence, and Teitelbaum were automating the labor of a far larger number of human computers who had preceded them, including themselves. As Munn's source for the quotation, Schwartz (2019), puts it: "Adele and Herman Goldstine, a married couple who led the human computing operations at BRL [the US Army Ballistic Research Laboratory], suggested that the most adept mathematical minds from their group should carry out the task. Together they selected six women [. . .] to be promoted from human computers to machine operators."[8]

These six human computers retained employment by becoming computer programmers, while other human computers became unemployed. As Grier (2007) put it, by the time of the Moore School Lectures organized in 1946 as a response to public interest in ENIAC, "the human computer was already starting to fade from memory. Most of the wartime computing groups had been shut down, reduced to a small remnant, or replaced by punched card equipment. The major employer of human computers, the Applied Mathematics Panel, had finally ceased operations after one last meeting" (Grier 2007, 288). Grier's final chapters chronicled the final demise in 1949 of the Mathematical Tables Project, a substantial and significant WPA-era human computing organization, and that of the Institute for Numerical Analysis subsequently directed at UCLA by Gertrude Blanch, the Mathematical Tables Project's technical leader. Of the former, Grier argued that "the death blow for the group was delivered by John von Neumann, who [. . .] suggested 'that a group of this type will become obsolescent when automatic devices become widely distributed'" (Grier 2007, 294). On Blanch's final day at the Mathematical Tables Project office in New York in 1948, Grier relates,

the computers took a break from their linear programming calculations and gave her a farewell party. Surrounded by adding machines and piles of paper, they shared one last plate of food with Blanch, offered her a parting gift, and left her a card expressing their best

wishes for her future. The computers came from the working-class
neighborhoods around New York City: Brooklyn, Yonkers, Jersey City,
the Bronx, Harlem, and the Lower East Side. A few of them would be
offered positions in Washington, D.C., but most would soon be looking for work in New York. None of them would be following Blanch.
(Grier 2007, 299)

Regarding Blanch's next position, at the UCLA Institute for Numerical
Analysis, Grier observed that by the time Blanch resigned this position
in 1953, "the engineers and researchers who needed computing services
had begun acquiring calculating equipment of their own [. . .] The larger
customers had ordered IBM's first electronic computer, the Model 701,
which had arrived on the market a year before. Fully six of the first
twelve customers for the 701 were companies that had once requested
computations from the Institute for Numerical Analysis" (Grier 2007,
310–11). While Blanch went on to a successful career with the US Air
Force Aerospace Research Laboratories, by the time of her retirement
in 1967, the National Institute of Standards and Technology had closed
its Computational Laboratory, one of the last human computing organizations, and while "a few businesses, such as insurance firms and petro-
leum refiners, retained small staffs of calculating assistants [. . .] these,
too, were being replaced with IBM 370s, DEC PDP-8s, Burroughs B-
6500s, and other computers that were powered by electricity and sported
numerical names" (Grier 2007, 318).

Of course, it didn't stop there. The kinetic momentum of a specific
history that herded Blanch and other human computers from one interesting, yet frangible way of making a living to the next, remaining one
step ahead of a wave of automation, was transferred to the professional
activity of programming itself, the occupation of the best-known surviving human computers, those who didn't simply vanish into the remascu-
linized postwar labor market and general social order of the 1950s. And
that momentum can carry us to the present day.

———

Coupland's *Microserfs*, mentioned above, is presented as the journal of
Daniel, a twenty-six-year-old software tester at Microsoft who leaves for
a Silicon Valley startup called Oop! Focused on the dynamics of the relationship between corporate giants like IBM and Microsoft on the one

hand and the startup culture of the West Coast, on the other, it is among the more light-hearted of the Silicon Valley novels that have anything substantive to say about the relationships among major divisions of the knowledge work force in the US tech industry and the mediation of those relationships by automation. In some ways, what is most interesting about *Microserfs* is that it is set in 1993 and published in 1995: more than anything else, it is a portrait of that tech industry between boom cycles. The internet stock frenzy of the dot-com boom that followed Netscape's IPO began only in the novel's year of publication, which lends a special tint to the complaints of "the first generation of Microsoft employees faced with reduced stock options and, for that matter, plateauing stock prices" (Coupland 2008, 16) and their "fear that the growth years will never return again" (38). (On December 31, 1992, Microsoft stock opened at $2.69 per share; seven years later, on December 31, 1999, it closed at $58.38.)

Though as a "bug checker," Daniel is part of the low-status segment of the Dev side of Microsoft's workforce, he is as soaked in its antipathies and self-regard as any of the application programmers. Of his "boss" (quotation marks in the original) Shaw, Daniel observes that "one grudgingly has to respect someone who's fortysomething and still in computers. [. . .] My only problem with Shaw is that he became a manager and stopped coding. [. . .] Managers either code or don't code, and it seems there are a lot more noncoding managers these days. Shades of IBM" (Coupland 2008, 33). The freewheeling temperamental irony of Daniel's reflection (such as it is) on his job is funneled into structural ironies like his own father's sudden loss of his managerial position at IBM. "We started talking more about all of the fiftysomethings being dumped out of the economy by downsizing. No one knows what to do with these people" (Coupland 2008, 23). It is not only managers, of course, who in abandoning the creative act of writing new user-facing application code have made themselves disposable, but the entire IT maintenance operation, whom Daniel regards as gray and invisible, both agents and casualties of automation. Oop!, where Daniel assumes the role of lead Windows developer for the product's cross-platform class library, is a classic startup, all Dev and no Ops. "We inhabit our workstations daily for a minimum of 12 hours" (Coupland 2008, 119), he observes. "Life has become coding madness all over again—except *this* time we're killing ourselves for our*selves*, instead of some huge company" (Coupland 2008, 135).

In Bronson's *The First $20 Million Is Always the Hardest*, Andy Caspar, a low-status software tester in the marketing department of the chip manufacturer Omega Logic, takes a position at La Honda Research Center, an elite nonprofit R&D institute providing services to Omega from its secluded and unassuming campus in the foothills near Stanford University. Assigned to an unprivileged back-burner project, the development of a cheap (three-hundred-dollar) computer, Caspar hatches a vision of a "VWPC" or people's PC, an explicit appeal to the engineering of the Volkswagen Type 1 economy car, otherwise known as the "Beetle" and "Bug." When Caspar and his team leave La Honda to launch a startup, such noble intentions are not the only thing driving them. Equally powerful is the romance of the startup as disencumbrance, or freedom from mere maintenance:

> Every self-respecting young man in Silicon Valley had considered the fantasy at one time or another. You could be pulling down a six-figure salary at Apple or Oracle [. . .] you could have two patents registered under your own name [. . .] you could have a division of programmers under your command [. . .] and it didn't count for *nothing* unless you had taken a crack at the big spin. You had to know: "Could I do it?" (Bronson 1998, 145)

And yet despite the premium placed on innovation, in an effort to conserve development costs, the VWPC team is driven to virtualization, in the automated simulation of the VWPC operating system on already existing computers. While the "Hypnotizer," a software virtualization environment that could be downloaded via a web browser and used to run otherwise platform-agnostic software, enables the production of internet-distributed software applications, it is a new affordance rather than a new product as such, entailing a commitment to building network infrastructure, beginning with "a firewall computer devoted to security; a router computer to direct traffic; a tower of modems to handle all the incoming calls; a high-volume internet link from the phone company" (Bronson 1998, 184). When in the novel's final segment, the VWPC project's rivals at their former employer, La Honda, examine the Hypnotizer's code, written in a domain-specific custom programming language devised for its purpose, the compactness that enables it to be directly interpreted, rather than compiled to object code, is an expression of the inventive agility of the startup culture in its freedom from the burdens of higher management and "human resources" as well as from IT operations; but the irony is that technically, the project's innovation consists in it serving as a platform for something (for potentially anything) else.

Whereas the setting of Bronson's novel, like Coupland's *Microserfs*, is in the few years immediately preceding the runup to the dot-com boom, Ullman's *The Bug* (2003) is set in 2000, the year of the dot-com crash, which in a brief yet emphatic moment of punctuation inscribed the contradictions of what Schiller calls "digital capitalism" into the surfaces of public life for the first time. Roberta Walton, its narrator, is a forty-eight-year-old consultant living off stock options as the crash begins, prompting a reverie that reaches back to the beginning of her tech industry career, sixteen years earlier, as the recipient of a doctoral degree in linguistics who, unable to find a permanent teaching position, took a job as a software tester at a startup. In the novel's fourteenth chapter, titled "The Night Systems Administrator," a romance between Walton's former colleague Ethan Levin, the developer whose pursuit of the elusive user interface bug UI–1017 provides the novel's core drama, and Ute Weiss, the sysadmin who fixes the bug (Ethan believes) by hauling his test server hardware away, is an allegorization of the interdependence of development and operations in the world before the Web. Its culmination in perfunctory, businesslike, mutually self-interested sex is not an indictment of this relationship, the story of which is folded into the metanarrative of economic violence that makes the dot-com crash an economic generalization of the inherent fragility and errancy of software.

The DevOps novel of the post-2008 moment is *The Phoenix Project: A Novel about IT, DevOps, and Helping Your Business Win*. Its authors, three IT consultants (as well as a founder of Tripwire, Inc., in the case of Kim, and a Purdue University professor of computer science, in the case of Spafford), present it as a portrait of "a core, chronic conflict between Development and IT Operations [that] preordains failure for the entire IT organization." "We now know," they argue, "that it is possible to break this conflict [. . .] the proof is that high-performing organizations such as Amazon, Google, Twitter, Etsy, and Netflix are adopting a set of techniques that we now call DevOps, and they are routinely deploying hundreds or even thousands of production changes per day, while preserving world-class reliability, stability, and security. By instituting a set of cultural norms, processes, and practices, these high performers are achieving breathtaking performance" (Kim, Behr, and Spafford 2014, 347).

Unlike *Microserfs, The First $20 Million Is Always the Hardest*, or *The Bug*, which are probably best described as literary portraits of elite white-collar work, *The Phoenix Project* is a business novel, composed programmatically for a readership of managers and other executives, and its principal creative motivation is to instruct in the pursuit of competitive advantage. But the conflict it portrays is deeply cultural, and in the battle of the makers and the maintainers that renders a fault line in the post-2008 US economy and its media reflection, defined by desperate innovation less as a given and more as a talisman against degrowth than at any time in recent history, the novel sides aggressively with the grunts against the glamorous. Its portrait of developers (application programmers, software engineers) is brutal. "Show me a developer who isn't crashing production systems," Bill Palmer, the novel's narrator and newly promoted VP of IT Operations at Parts Unlimited, an unremarkable and uninspired manufacturing and retail company, insists early in its second chapter, "and I'll show you one who can't fog a mirror. Or more likely, is on vacation" (Kim, Behr, and Spafford 2014, 28). "They're often carelessly breaking things and then disappearing, leaving Operations to clean up the mess" (39). In the words of the cantankerous Wes Davis, the company's Director of Distributed Technology Operations who reports to Palmer in his new role, Ops always needs a fixer, someone "to tell those idiotic developers how things work in the real world and what type of things keep breaking in production. The irony, of course, is that he can't tell the developers, because he's too busy repairing the things that are already broken" (Kim, Behr, and Spafford 2014, 56).

The Phoenix Project begins with the full text of an analyst's rating note reporting that Parts Unlimited's CEO has stepped down under pressure as the company's stock falls to a three-year low. Losing market share to a mysterious "arch rival, famous for its ability to anticipate and instantly react to customer needs" (Kim, Behr, and Spafford 2014, n.p.) and under pressure from institutional investors, Parts Unlimited has promised to upgrade its retail and e-commerce systems, but the development of this system, given the name "Phoenix," is years behind schedule. With a sell rating on its stock and a target share price of $8.00 against a current share price of $13.00, the heat is on. The analyst's sources claim that the board has given the executive team six months to turn things around.

In its mission to support deployment of Phoenix, IT Operations is hobbled by the volume of work "off the books" requested by executives who "go to their favorite IT person" (Kim, Behr, and Spafford 2014, 71) with strictly local problems and needs. These substantial, yet undocumented internal projects represent the maintenance function of Ops in the strictest sense, without which the company could not function at all: routine server and database upgrades needed for stable and secure performance, for example, or virtualizing applications for convenience in access, rather than to introduce new features, affordances, and efficiencies. The double-booking that sees Ops tasked also with supporting innovation in infrastructure is simply untenable, and the novel's happy path to management of the unmanageable runs alongside the detailed misery of its protagonists: confident, comfortable, and well-compensated knowledge workers whose working conditions are privileged and yet unbounded, grounded in the autonomy of their expertise, yet fundamentally confined in its application, and always on the precipice of self-automation, of inventing themselves out of their jobs. "Let me guess," Erik Reid, the powerful, purportedly brilliant company eccentric and self-styled guru tells Palmer during one of the latter's several crises of confidence. "You're going to say that IT is pure knowledge work, and so therefore, all your work is like that of an artisan. Therefore, there's no place for standardization, documented work procedures, and all that high-falutin' 'rigor and discipline' that you claimed to hold so near and dear" (Kim, Behr, and Spafford 2014, 91).

To be sure, Dev has its own problems. Chris Allers, VP of Application Development, Palmer tells us, "is constantly asked to deliver more features and do it in less time, with less money"; "to help with Phoenix, his team has grown by fifty people in the last two years, many through offshore development shops" (Kim, Behr, and Spafford 2014, 50). The generalization of network infrastructure that deskilled the sysadmin made it easier to relocate programming labor, as well (this being the novel's single, entirely oblique reference to outsourcing hubs like Bangalore, India). When Palmer meets him for a drinking lunch at a particularly harrowing moment, when Parts Unlimited's acting CIO has proposed outsourcing all IT Ops and the company's board has proposed splitting it up and selling it off, Allers, who has always struck Palmer as an archetype of Dev confidence, arrogance, and enjoyment of work, surprises Palmer with an admission. "Maybe my group being outsourced wouldn't be the worst thing in the world. [. . .] I used to love this work, but it's gotten so much more difficult over the last ten years. Technology keeps

changing faster and faster, and it's nearly impossible to keep up anymore
[. . .] It's crazy what programmers, and even managers like me, have to
learn every couple of years" (Kim, Behr, and Spafford 2014, 149–50). ("I
laugh sympathetically," Palmer reflects. "I chose to be in the technology
backwaters. I was happy there. That is, until Steve [Parts Unlimited's
acting CIO] threw me back into the big, shark-infested pool" [Kim, Behr,
and Spafford 2014, 150]). Then there is the extent to which Dev is directly
tasked with compensating for the mismanagement of Ops. As Erik Reid
puts it: "IT Operations seems to have lodged itself in every major flow of
work, including the top company project. It has all the executives hop-
ping mad, and they're turning the screws on your Development guy to
do whatever it takes to get it into production" (Kim, Behr, and Spafford
2014, 86).

It is at this point that Reid assumes his role as Palmer's secret advi-
sor, escorting him on a whirlwind tour of postwar manufacturing process theory and project management philosophy. It begins with a visit
to Parts Unlimited's manufacturing plant MRP-8, where the mysterious
Reid is addressed with the title "Doctor" and granted access to a catwalk
over the plant floor that is otherwise off limits. And from this point forward, Reid's instruction of Palmer represents a fictionally dramatized
version of the content of the novel's seven nonfictional appendices, more
than thirty pages in total, in which Kim, who with Behr and Spafford
authored in 2013 the first of a series of widely circulated annual white
papers on the topic, lays out the theory and practice of DevOps as a defining extension of manufacturing principles into knowledge work.

In the fictional narrative, Reid tells Palmer that during the 1980s,
production at MRP-8 was reorganized according to three "scientifically-
grounded management movements" (Kim, Behr, and Spafford 2014, 89):
the Theory of Constraints; Lean production, otherwise known as the
Toyota Production System; and Total Quality Management. In the nonfictional appendix "Where DevOps Came From," we are told that DevOps is
aligned with "an astounding convergence of philosophical management
movements," including "Lean Startup, Innovation Culture, Toyota Kata,
Rugged Computing, and the Velocity community" (Kim, Behr, and Spafford 2014, 355). In the fictional narrative, Reid explicates the Theory of
Constraints as a description of the power of bottlenecks in a production
process: Eliyahu M. Goldratt, he tells Palmer, realized that "improvements made *anywhere besides the bottleneck* are an illusion [. . .] In our

case, our bottleneck was a heat treat oven, just like in Goldratt's novel *The Goal*" (Kim, Behr, and Spafford 2014, 90). In the nonfictional appendix "Further Reading," Kim tell us that he and his co-authors studied this 1984 business novel (full title *The Goal: A Process of Ongoing Improvement*), about a plant manager given ninety days to resolve a production emergency or face shutdown, "for nearly a decade" while planning *The Phoenix Project* (Kim, Behr, and Spafford 2014, 367).

The earlier novel, a bestseller reissued in a thirtieth anniversary edition in 2014 and followed by a graphic novel version three years later, is also set in the United States, but in an entirely different time and place, and it is the vicissitudes of economic history rather than the ontological difference between physical widgets and binary blobs that measures that distance. Goldratt's insistence, in his introduction to the first edition, that "the western world does not have to become a second or third rate manufacturing power" (Goldratt and Cox 2014, n.p.), the novel's setting in Bearington, a crumbling rust belt factory town that "has been losing major employers at the rate of about one a year ever since the mid-1970s" (Goldratt and Cox 2014, 13), indeed, the obsession with finding something, anything, to "beat the Japanese" (Goldratt and Cox 2014, 17) all contrast strongly with the unglamorous but sturdy economic security of enterprise IT in *The Phoenix Project,* set in an economic-historical nonmoment almost completely unmarked by the boom-and-bust cycles of the operations it represents. Palmer and Allers may be oppressed by the ongoing, unabated crisis of software engineering, by the ever-accelerating pace of technical change, and by overwork and its consequences for mind, body, and family life, but Parts Unlimited's troubles are process and personnel management troubles through and through, and the worst that will happen if the company is split up is that its elite employees will drive down the road to their next jobs. There is no larger economic history here, nothing even remotely close to the metanarrative of global overcapacity in the long 1970s whose shadow falls so audibly on *The Goal*—indeed, no deep and genuine economic anxiety whatsoever.

Kim, Behr, and Spafford present *The Phoenix Project* as "an homage to *The Goal*" that deliberately mirrors "most of the book structure and plot elements" (Kim, Behr, and Spafford 2014, 367) while adapting it to twenty-first-century knowledge work. "Mirroring" is the right word for *The Phoenix Project*'s technical reproduction of both Goldblatt's scenario and his craft, while adaptation stands not merely for Kim, Behr, and Spafford's active variations on the original context, but for those enacted (or exacted) by history itself. Nothing in *The Phoenix Project,* which as DevOps fiction takes software-driven automation for granted and moves straight to its management implications, corresponds to the soul-searching focused on industrial robotics by Jonah, the "manufacturing management scientist" who is Goldblatt's autofictional alter ego and who plays guru to Alex Rogo, a plant manager at the UniWare division of UniCo, as Reid does to Palmer in *The Phoenix Project.* When Jonah asks Rogo, "Your company is making thirty-six percent more money from your plant just from installing some robots?" (Goldratt and Cox 2014, 27), Rogo understands only in retrospect that the query was thoroughly Socratic: "I remember that I was the one who was smiling. I was the one who thought *he* didn't understand the realities of manufacturing. Now I feel like an idiot" (Goldratt and Cox 2014, 66). "Yes, Jonah," Rogo reflects, "you're right; productivity did not go up thirty-six percent just because we installed some robots. For that matter, did it go up at all? Are we making *any* more money because of the robots?" (Goldratt and Cox 2014, 66).

The eponymous goal, and its role in *The Goal* as a business novel, depends on this insight about automation, and it is discovered through a process of elimination in these Socratic exchanges with Jonah. Eventually, Rogo is led to the insight that as a manufacturing company UniCo's single goal, on which his career has come to depend with new urgency, cannot be the efficient supply of raw materials; it cannot be keeping up with production technology; it cannot be the procedurally efficient, low-cost, or high-quality production of finished goods; it cannot be robust sales, market share, the creation of jobs, or some other relatively abstract mode of economic power. In the end, it is the most decisively abstract (and colloquial) mode of economic power of all: the power to "make money" (Goldratt and Cox 2014, 41). Since nothing so crass is permissible in the muted but still legible California ideology of *The Phoenix Project,* the latter's structural analogy for *The Goal* rests on the absence of integration between manufacturing and sales, the two cultures of which Rogo understands as a central obstacle. Of Eddie, the plant's second-shift supervisor, Rogo reflects that his world "is one measured in terms of parts per hour, man-hours worked, numbers of orders filled [. . .] Net profit, ROI, cash flow—that's just headquarters talk to Eddie" (Goldratt and Cox 2014, 51). Bob Donovan, Rogo's production manager, articulates this perspective explicitly: "We're manufacturing. We've got nothing to do with sales; that's marketing" (Goldratt and Cox 2014, 73).

In other words, while human workforce relations survive the three-decade translation of *The Goal* to *The Phoenix Project,* in the parallelism aligning Dev vs. Ops with manufacturing vs. sales, both the agents and the objects of automation have changed. In the nonfictional appendix "Further Reading," Kim reminds us that *The Goal* identified an industrial robot in Rogo's plant as the first of the constraints in Goldratt's "Theory of Constraints (TOC) methodology," whereas in *The Phoenix Project* "the constraint was initially Brent" (Kim, Behr, and Spafford 2014, 368). Brent Geller, who holds the title of Lead Engineer, is the archetype of the indispensable but unmanageable techie, who retains entirely unformalized, undocumented knowledge of the company's systems in his head and handles its emergencies largely on his own, as if by magic. "Brent, Brent, Brent!" Palmer exclaims early in his struggle to gain control of the situation. "Can't we do anything without him?" But Wes Davis's reply is: "He knows a lot about almost everything we have in this shop [. . .] Heck, the guy may know more about how this company works than I do" (Kim, Behr, and Spafford 2014, 70).

Other works specified in Kim's nonfictional appendix "Further Reading" include Mike Rother's *Toyota Kata: Managing People for Improvement, Adaptiveness and Superior Results* (2009) along with David H. Anderson's *Kanban: Successful Evolutionary Change for Your Technology Business* (2010), Jim Benson and Tonianne DeMaria Barry's *Personal Kanban: Mapping Work/Navigating Life* (2011), and two books on contemporary software project management, Michael T. Nygard's *Release It! Design and Deploy Production-Ready Software* (2007) and Jez Humble and David Farley's *Continuous Delivery: Reliable Software Releases through Build, Test, and Deployment Automation* (2010). But the second most important model for *The Phoenix Project,* according to Kim, is Patrick Lencioni's *The Five Dysfunctions of a Team: A Leadership Fable* (2002), one in an extended series of novel-length business fictions including *The Five Temptations of a CEO* (1998), *The Four Obsessions of an Extraordinary Executive* (2000), *Death by Meeting* (2004), and *The Ideal Team Player* (2016). "When I think about the long, bitter intertribal warfare that has existed between Development and IT Operations, as well as between IT and 'the business'," Kim reflects, "I suspect that we will very much need the lessons of Mr. Lencioni to achieve the DevOps ideal" (Kim, Behr, and

Spafford 2014, 369). The shadow this casts over *The Phoenix Project*'s sunny narratological disposition and its unequivocally happy ending suggests that the surface tension of DevOps fiction can support the arrangement of a variety of players. When Wes Davis suggests that Brent "may know more about how this company works than I do," Palmer's exasperation is clear: "You're a senior manager. This should be as unacceptable to you as it is to me!" (Kim, Behr, and Spafford 2014, 70). But, with no way to access what Brent knows, much less to share it, and thus no way to replace him, Palmer is forced to protect Brent's time and ability to focus on impediments to Phoenix's launch, on which the company's future depends.

Subsequently, after another conference with Reid, Palmer identifies Brent as an explicitly destructive constraint in his structural exclusion from management: "Look, it's rocket scientists like him that often cause the problem in the first place" (Kim, Behr, and Spafford 2014, 171). In Kim, Behr, and Spafford's translation of Goldratt's theory of constraints, the characterological and narratological replacement of an industrial robot with a human worker whose understanding of automated systems is private, unshared, deeply idiosyncratic, and embedded in the fundamental technical operations on which the company depends seems a significant transposition. Is this merely a compositional artifact of the earlier novel's focus on industrial manufacturing, and its situation preceding the software booms of the twenty-first century? That is for (more) time to tell. It may well be that the difference that makes a difference, here, lies in the logic of automation, which proceeds by replacement in the one instance—semi-skilled human laborers replaced by robots, with the "surprising" result that no gains in productive efficiency follow automatically and spontaneously—and recursion in the other, as the quiddities of craftspersonship are reinjected into mature postindustrialization by the temporary but real, and profoundly *autological* indispensability of the highly skilled, significantly autonomous para-managerial technical worker.

ACKNOWLEDGMENTS

I thank Erica Wetter at Stanford University Press for her encouragement of this project and her patience regarding my pace. I also thank Caroline McKusick for her support and assistance. I was fortunate to receive reports by peer reviewers who considered the manuscript with care and grasped its purpose well enough to provide me with productive responses and suggestions for revision.

I thank Duke University Press, Johns Hopkins University Press, and Springer Nature B.V. for permission to draw on earlier versions of the material in this book. I thank Massachusetts Institute of Technology, IBM Corporation, the International Federation for Information Processing, and Oracle and its affiliates for permission to reprint the figures that appear in the book.

NOTES

Introduction

1. A passage in this introduction originally appeared in "Program Resumed," an interview published by the Johns Hopkins University Press blog, *Newsroom,* on March 28, 2018.

2. See Cortada (2004), Cortada (2006), and Cortada (2008).

3. For the claim that 220 billion lines of Cobol are still in use today, see "COBOL Blues" (n.d.). See also Cassel (2017) and "Did COBOL Have 250 Billion Lines of Code and 1 Million Programmers, as Late as 2009?" (2017).

4. See Ford (2015), Nguyen (2015), and Somaiya (2015).

5. Elon Musk, whose opinions still carry weight even if at this point, they're the equivalent of a stopped clock that tells the right time twice a day, recently opined that "too much talent is chasing software" in the tech industry and more generally, calling it a "major misallocation of capital." See Tangermann (2023).

6. For a journalistic perspective on this point, see Merchant (2018). As I acknowledge at other points in this book, mine is a generalizing and generally looser usage of the term "recursion" than a strictly technical perspective might permit. That is deliberate, not accidental.

7. See especially Fuchs (2014), Fuchs and Mosco (2016), and Fuchs (2016), but also Terranova (2004), Bellucci (2005), Scholz (2013), and Scholz (2017), among others.

8. See Light (1999), Campbell-Kelly (2004), Grier (2007), Ensmenger (2010),

Abbate (2012), and Hicks (2017), among others (and before them, Kraft [1977] and Greenbaum [1979]). For a useful survey and evaluation of what the author calls the "sociology-of-work literature in this field," including studies published by Fred Block, Alvin Gouldner, Larry Hirschhorn, Rosabeth Kanter, Michael Piore and Charles Sabel, Alain Touraine, and Shoshanna Zuboff, whose early work optimistically counterposed "informating" to "automating," see Burris (1993, 16–18).

9. See Williams and Srnicek (2013).

10. As examples of work in these areas, see Benjamin (2019), Berry (2011), Bratton (2015), Chun (2013), Evens (2015), Franklin (2015), Frabetti (2015), Golumbia (2009), Golumbia (2016), Hui (2016), Noble (2018), Sack (2019).

11. See Mackenzie (2006).

12. A partial exception is Montfort et al. (2013), seven of the eleven main chapters of which focus on the BASIC programming language. Only one of these seven chapters could reasonably be called a study of the BASIC language, however, with the remaining six devoted to explaining very simple command sequences and very brief programs to readers who are assumed to have no knowledge either of BASIC or any other programming language. The book's eighth main chapter, titled simply "BASIC," does discuss language design and syntax variation in some detail, but is otherwise given over to reviewing BASIC's implementation and usage history, again for a reader assumed to lack elementary knowledge of the subject. As of this writing, *10 PRINT CHR$(205.5+RND(1));:GOTO 10* and Marino (2020) are the only book-length publications to have emerged from "critical code studies," an undertaking that Marino (2006) attempted, laudably if probably unsuccessfully, to distinguish from a more media-oriented "software studies" almost two decades ago. As for the more general "digital humanities" movement, its inaccurate but fundamental premise, that humanities scholars are uniformly newbies to technical aptitude and training in programming, leaves it unfit for advanced research in this area. Those few self-identified digital humanities scholars who *do* write programs in R and Python, for example, for their work are applying these programming languages as tools for research automation, not studying them as research objects in themselves. In short, when digital humanities scholars were presented with the opportunity to study programming languages as cultural artifacts relatively closely aligned with their training and expertise, they chose instead to become the world's most incompetent and bumbling research programmers. See Swafford (2015), Da (2019), and Levy-Eichel and Scheinerman (2022).

13. See Sammet (1969), Knuth and Pardo (1976), Wexelblat (1981), Bergin and Gibson (1996), Bergin (2007).

14. The "politico-social history of Algol" promised by Bemer (1969), for example, turns out to be a bibliography with abridged extracts from various primary sources (letters, meeting minutes, committee resolutions, and so on), many relating to the famously fractious negotiations of the specification of Algol 60 in particular. It is that history of conflict to which the term "politico-social" pre-

sumably refers; still, this document is entirely descriptive and offers no analysis whatsoever.

15. See Light (1999), Grier (2007), Abbate (2000), Abbate (2012), Ensmenger (2010), McIlwain (2020), Shetterly (2016), Hicks (2017).

16. See Lennon (2013).

17. See Ames (2019).

18. See Felleisen et al. (2018).

19. See Lopes (2014), Lopes (2021), Croll (2015).

20. See Raley (2002).

21. Another similarly approving citation, equally difficult to reconcile with Marino's rejection of the conflation of language and code, is that of Geoff Cox and Alex McLean's programmatic pastiche of code with speech (Marino 2020, 12). Again, for me this is a nonstarter. See Cox and McLean (2013).

22. See Said (2004), Bové (2009), and Mufti (2010).

23. "Social historians," Haigh (2014) observes, "have done a great job examining the history of ideas like 'freedom' and 'progress,' which have been claimed and shaped in different ways by different groups over time. In the history of the past 60 years ideas like 'information' and 'digital' have been similarly powerful, and deserve similar scrutiny. If I was a 'digital historian,' whose own professional identity and career prospects came from evangelizing for 'the digital,' could I still do that work?" (Haigh 2014, 28). For a sociological analysis of catachrestic usages of the word "digital" more generally, see Sweeney (2015).

24. On the initial reception of ChatGPT in higher education, see the summary articles by McMurtrie (2022) and Warner (2022).

25. In August 2023, only two years after acquiring coding bootcamp Kenzie Academy, Southern New Hampshire University decided to close the program. A spokesperson suggested that the potential impact of LLM-powered automation, on either demand for such programs or their outcomes (or both), was a factor: "Lopez [. . .] said the 'exponential' adoption of artificial intelligence played a role in the decision to end Kenzie. 'The world is changing fast, and we, like many institutions, are thinking about the ways AI will impact education in profound ways that are just starting to take shape,' she said." See Coffey (2023).

26. The text quoted here was generated on March 7, 2023, in response to the prompt "How likely is it that adoption of Codex will reduce demand for human programmers and software engineers?"

Chapter One

1. The December 2010 update of the third edition of the *Oxford English Dictionary* included revised definitions for both "digital" and "analogue." The newly revised entry for "digital" includes a fifteenth-century usage, "designating a whole number less than ten," that had not appeared in the first edition (Holden 2010).

2. Ceruzzi (2003), 1. See also Abbate (2012) and Ensmenger (2010).

3. The diagram suggests that the architecture in question provides 511 index registers for memory storage. See "Preliminary Specifications: Programmed

Data Processor Model Three (PDP-3)" (1960): "PDP-3 is a stored program, general purpose digital computer. It is a single address, single instruction machine operating in parallel on 36 bit numbers. It features multiple step indirect addressing and indexing of addresses. The main memory makes 511 registers available as index registers."

4. The design of a discrete software compiler to perform such a task is an accomplishment for which Grace Murray Hopper is credited, around the same time.

5. See Adams and Gill (1954), 1-3, 1-5.

6. See, for example, Backus and Herrick (1954), 106.

7. This stabilizing and maturation of technical terminology represents a movement away from the borrowing and metaphorizing of linguistic terminology in which the term and concept of translation served as temporary historical placeholders. Of course, the translation of a literary text from one natural language to another presents insuperable obstacles to precision where any mode of equivalence is concerned (a dynamic whose description is central to translation theory in the humanities and a central topic in my own previous scholarship). This might be imagined as nullifying "translation" altogether as a descriptor of the process of decompiling assembler back into C or C++, for example, as a process designed to produce identical code. Yet it is widely acknowledged that highly technical or otherwise specialized prose in natural languages, rendered using a specialized lexicon deliberately designed to control polysemy, ambiguity, and context dependence in a specifically expert use case, can indeed be translated rather precisely into a second natural language and back again, without any enormous loss.

8. To suggest that such usage is firmly settled today, and that the lexicon emerging in the 1950s was less settled and that it overlapped with terms from everyday linguistic usage, is not to suggest that some familiar technical terms had not already emerged by the 1950s. Grace M. Hopper (1956), for example, defined and distinguished clearly and carefully between *converter, interpreter, generator, assembler,* and *compiler* as software components operating on *pseudocode* and producing *computer code*, and between *converting, interpreting, generating, assembling* and *compiling* as primary computational processes. It is notable, however, that at the same time, Hopper made free use of the translation metaphor and the word "translation" to describe all of the latter processes—something one would find neither professional software developers nor academic computer scientists doing today.

9. See Nofre, Priestley, and Alberts (2014), 47–48.

10. On this point, see Priestley (2010), 2.

11. Readers will recognize the unmistakable echoes here of Vannevar Bush's "As We May Think" and Warren Weaver's 1949 memo on machine translation. See Bush (1945), Weaver (1949), and Weaver (1955).

12. See Pflüger (2002), 125–26. See also S. Gorn (1960): "From the moment that von Neumann first suggested that the instructions and data should share

the same storage in a machine there has been a growing realization that general purpose digital computers are linguistic mechanisms" (117).

13. See Abbate (2012), 85, 102–11.

14. See US Bureau of Labor Statistics (USBLS) (2014), (2015), (2020), (2022). As of this writing, the growth forecast provided by the USBLS for the category "Software Developers, Quality Assurance Analysts, and Testers" is robustly positive, as is that for the category "Web Developers and Digital Designers." At some level, these differences are related to how the USBLS defines a "computer programmer," a "software developer," and a "web developer" as distinct occupations. To the best of my knowledge, the USBLS does not make publicly available any information about how these determinations are made. For most people today, a "computer programmer," a "software developer," and a "web developer" really aren't clearly distinguishable occupations; at the very least, most people would say that a "computer programmer" and a "software developer" are pretty much exactly the same thing. Nonetheless, it remains interesting, if not necessarily counterintuitive, that job growth in *any* of these three occupations might now be consistently forecast as negative. Possibly pertinent, as well, is that as of this writing, the USBLS's numerical forecasts and their accompanying prose explanations have not yet begun to absorb the impact of very recent technology industry layoffs—or that of the public releases, at exactly the same time, of a new generation of highly sophisticated automation products created by OpenAI, Microsoft, Google, and others.

15. Aronowitz and DiFazio (1994), Rifkin (1996), and Uchitelle and Kleinfeld (1996), as cited in Bix (2002), 300.

Chapter Two

1. An earlier version of this chapter, which also included passages from the introduction, appeared with the same title in *Cultural Politics* 14.3 (Duke University Press, 2018): 372–394. © 2018 by Duke University Press.

2. Daniel Punday gently resists both Espen Aarseth's and N. Katherine Hayles's critical investments in code "as a kind of *other* to natural language, a form of writing without a real audience," reminding his readers that "code is written not just for the computer but also for programmers to keep track of the design of the software for debugging, updates, and so on" (Punday 2015, 129). Presumably Punday has the distinction between program code and program comments in mind, though he does not say so. In a similar context, Federica Frabetti has rejected Hayles's imagination of code as writing's successor, acquiring "a much stronger connection to materiality than speech and writing" in its progressive-historical priority (Frabetti 2015, 43). Frabetti observes that the call for an analytic turn to materiality which Hayles grounds on that claim is "quite belated" and "results from a misreading of the poststructuralist tradition" it takes as a foil (Frabetti 2015, 51). I endorse these critiques of a replacement theology distinguishing natural language and code as both incommensurable *and* historically sequential, found in its most aggressive form in the work of Friedrich

A. Kittler, the allegorical substrate of which is an obsessive deposing of literary language. However, I see little value in conflating the apples of natural language with the oranges of code, and find its enactment tendentious where the benefits are obscure. I cannot follow Justin Joque, for example, in an otherwise exceptionally rich and rewarding essay, in suggesting that a programmer's "creative" writing of comments alongside code (and enjoyment of latitude in parameter naming and overall program design) elides their difference, if not literally then in a productively imaginative way (Joque 2016, 339). I could not agree more with Joque on the importance of studying programming languages, but we ought to treat them as what they are: technologies of automation, programmatically (but routinely) constructed to extinguish human agency. We can't say we're relitigating old questions, here, while the relevant technical competencies are so rare.

3. See Grafton (1999), Bush (1945), and Landow (2006), 119–22, 182–84, 279.

4. See Genette (1987), 20 and following, especially 35 and following ("Composition, tirages"); also 316 and following ("L'épitexte public") and 341 and following ("L'épitexte privé"); Genette (1997), 16, and following, especially 33 ("Typesetting, printings"); also 344 and following ("The public epitext") and 371 and following ("The private epitext").

5. See Nelson (1965) and "Transclusion: Fixing Electronic Literature" (2007).

6. In objecting to what I praise here as Etzkorn and colleagues' unobfuscated distinction between nonexecutable and executable program text, one might point to an artifact like the doctest module included in the standard library (that is, all the code packaged with the standard implementation, rather than installed separately) of the Python programming language. doctest provides the verification, through execution, of appropriately formatted illustrative code snippets included within the very specific type of comment block that Python programmers call "docstrings" (see the discussion of "documentation comments" or "documentation strings," below). But this execution of code snippets so to speak "within" comments does not somehow transform comment text itself into executable text; that is, it is not execution *of the comments* themselves and as such. Rather, a library like doctest must first parse nonexecutable comment text in order to locate and then extract code snippets before evaluating them as code in an environment created for that purpose. Both the provision and the use of such "executable docstrings" are, at least for the moment, relatively marginal in programming practice: most such comments are like others, written to be read along with the code they document, or nondestructively extracted into separate reference documents.

7. "Tooling" is a currently widely used term for software applications used exclusively by software developers, including not only integrated development environments (IDEs) and other applications that assist in generating, compiling, packaging, and distributing code and documentation, but also applications for formatting, testing, debugging, refactoring, continuous integration, issue and bug tracking, version control, performance analysis, disassembly, and other tasks.

8. On Short Code as the first implemented higher-level programming language, see Sammet (1969), 129–30, and Donald E. Knuth (2003), 19–22. See also Mauchly (1988), which is included in Schmitt (1988), 17–18.

9. On Speedcoding, see Sammet (1969), 130–31, and Donald E. Knuth (2003), 41–42. See also John W. Backus and Herrick (1954), J. W. Backus (1954), and "IBM Speedcoding System (from IBM Manual)" (1983). On the Laning and Zierler system, see Sammet (1969), 131–32, and Donald E. Knuth (2003), 45–49. See also Adams and Laning (1954), and Laning and Zierler (1954).

10. See *The Fortran Automatic Coding System for the IBM 704: Programmer's Reference Manual* (1956). John Backus, director of the IBM team that implemented Fortran I, credits David Sayre with the composition of the *Programmer's Reference Manual*. See J. Backus (1981), 36. Of Fortran I's provision for comments, Donald Knuth (2003) remarks that "no programming language designer had thought to do this before! (Assembly languages had comment cards, but programs in higher-level languages were generally felt to be self-explanatory)" (71).

11. Algol 58 comment syntax (using the reserved word *comment*) is specified in Perlis and Samelson (1958a) section II.5.v, "Comment declarations" (p. 394); Perlis and Samelson (1958b) section II.E.5, "Comment declarations" (p. 19); and Perlis and Samelson (1959) section II.5.v, "Comment declarations" (55). Comment syntax (using the reserved word *NOTE*) in Cobol 60 is specified in "COBOL: Initial Specifications for a Common Business Oriented Language" (1960) section V.6.2, "Verb formats" (p. V-31). There is no mention of Lisp I program comments in J. McCarthy et al. (1960). In John McCarthy et al. (1962), section VI.6.1, "Preparing a Card Deck," and Appendix E, "Overlord–The Monitor," a Lisp 1.5 program comment field is specified as beginning after column 16 of a punched sheet or card. See John McCarthy et al. (1962), 31, 81.

12. "C family" refers to programming languages whose syntax is recognizably partly or entirely derived from that of the C language created in 1972 by Dennis Ritchie. These include C++, C+, Objective C, Java, JavaScript, PHP, Go, and Rust, among others.

13. The most influential disseminator of the concept of structured programming was Edsger W. Dijkstra; see in particular Dijkstra (1968).

14. See Strunk Jr. (1920), 5, and Strunk Jr. and White (2009), 8.

15. The phrase "software crisis" refers to the conditions leading up to the introduction of structured programming (as discusssed above), but also to struggles with the management of large programming projects like IBM's OS/360 operating system. On the latter, see Brooks (1995). For an assessment of the progress or lack of progress made in software project management since the 1960s and 1970s, see Rosenberg (2008).

16. See Goodliffe (2007), 25, 59.

17. On the language metaphor in early computing, see Nofre, Priestley, and Alberts (2014).

18. The history of programming environments as we know them today begins with the so-called pipeline or toolchain including an assembler, compiler, and

linker, among other distinct but co-operating programmers' software tools that translate and integrate program text, producing a binary executable object (that is, the "app" on whose icon one clicks with a mouse or touches with a finger, to launch it). It includes so-called bytecode compiling for virtual machines, permitting the writing of programs that will run on different hardware processors and hardware memory configurations, rather than on only one computer model at a time. It includes interactive REPL (Read-Evaluate-Print-Loop) interpreters, a programmers' software tool used for exploratory stepwise programming (an alternative to writing program sections or even entire programs without knowing if they will run). Finally, it also includes so-called live coding environments, in which program code is continually reloaded and executed as it is being written and edited.

19. Possibly oddly, given his rejection of the narrowly conceived materialism of medium-specific approaches to computing, Punday's discussions of the work of Alan Kay are focused exclusively on the hardware design of Kay's Dynabook concept computer, with no mention of the Smalltalk system it was designed to support. See Punday (2015), 5, 23–24, 89.

20. The implementation referred to had been created by the British Royal Signals and Radar Establishment (RSRE). One difference between the two versions of Lindsay's essay concerns the abbreviated comment syntax character, **c** in the 1968 version but expanded to the hash symbol #, the character ¢, and the two-line abbreviation **co** by the time the Algol 68 Final Report had been published in December 1968. See Van Wijngaarden et al. (1968) and Van Wijngaarden (1969).

21. See Donald E. Knuth (1974), 670, and (1992a), 8.

22. In what follows, I cite the version of the essay "Literate Programming" reprinted in the 1992 book (Donald E. Knuth 1992b).

23. See Bentley, Knuth, and McIlroy (1986, 478). McIlroy's assessment of Knuth's literate program for tabulating the most commonly occurring words in a text file, which prompted Knuth to name McIlroy "the world's first literary critic of a computer program" (Donald E. Knuth 1992b, xi), was reprinted in *Literate Programming*; see Donald E. Knuth (1992b), 168 and following.

24. See Perrier (2015).

25. See Holen (2004).

26. See Bishop (2012).

27. See tali713 (2009).

28. See Collins (2016) and Garry (2009).

29. See "Funny Things Seen in Source Code and Documentation" (2014).

30. See Boden (2016), also McKinley (2009).

31. See Boden (2016).

32. See Boden (2016).

33. See Croll (2014): "Great literature is the creation of open minds. As programmers our focus can be very narrow. [. . .] As programmers we're not encouraged to experiment [. . .] It's amazing how many of us are afraid to break the rules that we're given, even when no one else is ever going to see the code."

Chapter Three

1. An earlier version of this chapter appeared with the same title in *Philosophy & Technology* 34.1 (Springer Nature B.V., 2021): 13–32. © 2021 by © Springer Nature B.V.

2. See Russell (2014), 168–69; Braman (2011), 296–98; Abbate (2000), 73–74.

3. DEL stood for Decode-Encode Language, according to Crocker the first version of a host protocol language. See Reynolds and Postel (1987), 2.

4. See Abbate (2000), 74, and Reynolds and Postel (1987), 3.

5. See Reynolds and Postel (1987), 7.

6. For the most recent version of *The Jargon File*, see Raymond (2004b).

7. See, for example, Mother of Five (1937), "On Foo-Ism" (1938), and "Temperance Poll Shows 87 Percent of Voters Imbibe" (1938) (examples provided in "What Does 'foo' Mean?" [2011]).

8. See, for example, Holman (1938).]

9. See "Science: Foo-Fighter" (1945).

10. On "snafu" and "fubar," see Elkin (1946), "Fubar" (2002), Burns and Novick (2007), Blackmore (2009), and "Fubar, Adj." (2016).

11. See P. R. Samson (1959).

12. On the language metaphor in early computing, see Nofre, Priestley, and Alberts (2014). In practice, programming language specifications can vary widely, and in some cases a standardization document, reference document, or reference implementation serves in place of a specification. For examples representing the most widely used languages today, see "ISO/IEC 9899:2018 Information Technology—Programming Languages—C" (2018), "ISO/IEC 14882:2017 Programming Languages—C++" (2017), "The Python Language Reference" (2019), Gosling et al. (2019), and "Standard ECMA-262: ECMAScript® 2019 Language Specification, 10th Edition" (2019).

13. See, for example, Kernighan (1974), McConnell (1993), and Goodliffe (2007).

14. The latter name reflects the complexity of graphical user interface (GUI) application components written in a programming language that strictly enforces an object-oriented programming style, such as Java. See, for example, Chartier (2004), Grouchnikov (2007), and Ynda-Hummel (2013).

15. See, for example, Berkeley and Bobrow (1964), Hart and Levin (1964), and P. Samson (1966).

16. See Sebesta (2010), 208, 211.

17. See Sebesta (2010), 209–10.

18. See Watson (2005), 142–52.

19. See Wentworth and Flexner (1900), Wentworth and Flexner (1975), Chapman and Wentworth (1986), and Chapman, Kipfer, and Wentworth (1995).

20. See Backus (1959). For Davis and Putnam's use of "syntactic variable" see Davis and Putnam (1960). The phrase "metasyntactic classes" and "metasyntactic formulae" appear in "Untitled" (1961) and Knuth and Merner (1961), both in discussing the Algol 60 specification. For a catalog of other early examples of these and similar usages, see Bron (2017).

21. See, for example, Hewitt, Bishop, and Steiger (1973), Nicholls (1975), and Galley and Pfister (1979).

22. See Sachs (1976).

23. See Sachs (1976).

24. See "Welcome to the Canvas LMS API Documentation" (n.d.), "MediaWiki API:Main Page" (n.d.), "A Plone API" (n.d.), "Drupal API" (n.d.), "Amazon Simple Storage Service API Reference" (n.d.), "GitHub Developer: REST API V3" (n.d.).

25. See, for example, Knuth (1997).

26. For another relevant discussion in this form, see Copeland (2010). For another example of a complaint, see Reisner (2012).

27. See Kernighan (1974), McConnell (1993), and Goodliffe (2007), as cited previously.

28. The classification of programming languages includes imagined fourth and fifth generations; see Martin (1982).

Chapter Four

1. See Montfort et al. (2013).

2. Control system applications, however, might certainly be understood as neither primarily numeric nor primarily textual in character.

3. A Van Wijngaarden grammar is another formalism developed for (later) Algol specifications, some of which (for example, revisions of the Algol 68 report) devised simplified ways of writing or annotating it. Backus-Naur form was subsequently repurposed in the many variants of so-called Extended Backus-Naur form. The same can be said of the many implementations of a Van Wijngaarden grammar—to choose an example relevant in this context, the so-called Q-systems devised for the TAUM machine translation project at the Université de Montréal. On the substitution of "Naur" for "Normal," see Knuth (1964).

4. That COMIT demoted arithmetic does not mean it was not useful for symbol-processing tasks outside the domain of natural language: while natural-language text data represented as strings was the primary object of both the MT researcher (at least initially) and the linguist in general, COMIT was intentionally agnostic with respect to modes and domains of symbolic notation. J. L. Darlington, for example, described an application of COMIT to logic translation, the creation of formal representations of natural-language input sentences as a storage or intermediary format for machine translation or abstracting and paraphrasing (Darlington 1965, 66).

5. E. J. Desautels and D. K. Smith suggest that Snobol syntax was derived from the notation used by A. A. Markov in *Theory of Algorithms* (1954). However, none of the many essays and reflections by Snobol's creators mentions Markov's book as a design source. See Desautels and Smith (1967), 419; Markov and Nagorny (1988).

6. See Sammet (1969), 382–85. See also Forte (1968), 161–62.

7. See Forte (1968), 158n3: "SNOBOL is sometimes described as a list-processor. This is incorrect."

8. Both operators could be used simultaneously, as in "foo" "o" :s(start):f(end).

9. "As examples of SNOBOL4's power," M. B. Emmer (1984) describes a version of Joseph Weizenbaum's conversation program ELIZA, implemented in "just 100 SNOBOL4 statements"; a recognizer for a Backus-Naur form grammar, in 30 statements; and a "program to copy a file and perform character substitutions [. . .] [which] is exactly 1 statement long" (12). Griswold reflected that "the most controversial aspect of the syntax of SNOBOL was its use of blanks. The decision not to have an explicit operator for concatenation, but instead just to use blanks to separate the strings to be concatenated, was motivated by a belief in the fundamental role of concatenation in a string-manipulation language. The model for this feature was simply the convention used in natural languages to separate words by blanks" (Ralph E. Griswold 1981, 617).

10. See http://www.snobol4.org/csnobol4/.

11. See Forte (1968) and R. E. Griswold, Poage, and Polonsky (1971). For my solutions to exercises, see http://bitfragment.net/exs-sno/.

12. See http://www.snobol4.org/dennis.heimbigner/s3/ and http://berstis .com/snobol4.htm.

13. See http://www.snobol4.org/.

14. See, for example, Paine (2010b); Paine (2010a); Jeffery et al. (2016).

15. "Synchronic" and "diachronic" are widely used terms in the study of linguistic and literary culture and of signification more generally. Where they appear together in current usage, their source is usually an English translation of *Course in General Linguistics* (*Cours de linguistique générale* [1916]), a volume originally assembled in French from notes on lectures by the Swiss linguist Ferdinand de Saussure (1857–1913).

16. "Proppian" here refers to the formalistic literary research of Vladimir Propp (1895–1970). The seven abstract character and thirty-one abstract plot structures into which Propp decomposed the Russian folktale has long made it attractive to programmers experimenting with computational plot and story generators. Many implementations are available.

17. "Dealings with intelligence organizations can be 'interesting.' The CIA gave us, verbally, a sequence of residential addresses in the Georgetown suburb of Washington, D.C. as places to deliver copies of SNOBOL4. When they called me, they gave no names, but only said 'you know who this is.' Somehow, I always did, although I was tempted on occasion to respond whimsically. When they wrote, which was rarely, their stationery had no letterhead—but it had an eagle watermark that was so prominent that it could be seen 10 feet away. On more than one occasion, a security-conscious organization degaussed the tapes we sent them on the way in, leaving the person who had requested SNOBOL with only a blank tape" (Griswold [1985], 4).

18. Alan C. Moore recalls that his "first experience with computers was in a class with Allen Forte at Yale in the mid 1960s dealing with computer applications to music theory. We were introduced to FORTRAN, SNOBOL and other languages" (Shannon 2003). Asked by his interviewer about the co-occurrence of

musical aptitude and aptitude for computer programming, Moore answers in the affirmative; asked about connections between computer programming and "the Humanities" more generally, Moore's answer is "Not as much in my opinion."

19. Brooks (1995) now stands at the head of an extensive literature chronicling struggle with the management of large programming projects. On the history of software engineering as a professional practice, see Frabetti (2015), 67–98, "Software as Material Inscription: The Beginnings of Software Engineering."

Chapter Five

1. Some readers may also have thought of the acquisition of Yahoo Inc. by Verizon Communications, announced on July 25, 2016, several months after an initial draft of this essay was completed.

2. See Brooks (1995).

3. An earlier version of this chapter appeared with the same title in *Configurations* 26.1 (Johns Hopkins University Press, 2018): 47–72. © 2018 by The Johns Hopkins University Press.

4. Designed by Bjarne Stroustrup while working at AT&T in the late 1970s, C++ was originally called "C with Classes," marking Stroustrup's intention to "superset" the C language (that is, to remain completely compatible with it) while also improving it. See Stroustrup (2016).

5. See, for example, McPeak and Wilton (2015), 2: "Perhaps this is a good place to dispel a widespread myth: JavaScript is not the script version of the Java language. In fact, although they share the same name, that's virtually all they do share. Particularly good news is that JavaScript is much, much easier to learn and use than Java." Even some classic books on JavaScript written for developers already expert in other languages, or newer books that address the same kind of reader, frame the issue similarly. In *JavaScript: The Definitive Guide,* widely considered an authoritative and comprehensive study of JavaScript, David Flanagan begins thus: "The name 'JavaScript' is actually somewhat misleading. Except for a superficial syntactic resemblance, JavaScript is completely different from the Java programming language" (Flanagan 2011, 1). In the introductory volume of a rigorous and well-received multivolume study of contemporary JavaScript, Kyle Simpson writes, "The name [JavaScript] is merely an accident of politics and marketing. The two languages are vastly different in many important ways. 'JavaScript' is as related to 'Java' as 'Carnival' is to 'Car'" (Simpson 2015, vii). Douglas Crockford reminds us of Java and JavaScript's historical concurrency and doesn't exaggerate their unrelatedness, but has little to say about the issue beyond one sentence: "When Java™ applets failed, JavaScript became the 'Language of the Web' by default" (Crockford 2008, 1).

6. In response to such claims, Java programmers often point to Java's leading position in the TIOBE Programming Community Index compiled by the software services provider TIOBE Software BV, or similar rankings aggregators like the PYPL Popularity of Programming Language Index—leaving unmentioned such rankings' historical "trend" indexes for Java's position, which are fre-

quently negative. See McMillan (2013) and Cassel (2016), both of whom argued at the time that TIOBE data itself showed Java's position "slipping" and "trending down" (McMillan also quoted Paul Jansen, managing director of TIOBE Software, as stating that "Java is falling down"). By 2022, the lead in the TIOBE index had been taken by Python, which remains there at the time of writing in 2023, having taken the lead in the PYPL index as far back as 2018. In any case, at any point in the history of a programming language past the point of its initial adoption, a language's reputation—as expressive or otherwise pleasant to use, as adaptable to ongoing hardware evolution, as usable in solving newer computational problems—may diverge from its market share or other measures of usage quite radically, if only because once they are in place, large industrial software infrastructures are kept operating for as long as possible. It is the incongruence of Java's reputation with its market share, today, that animates nonmeaningless if possibly glib comparisons of Java to Cobol, such as that made by Snyder (2007). The inclusion in Java version 9 of a REPL (Read-Evaluate-Print-Loop) feature for exploratory programming is a concession to the encroachment on Java's position of both scripting languages and newer functional programming languages, languages in both of which categories have offered REPL-type features—whose purpose and usage are fundamentally incompatible with Java-style object-oriented programming—for many decades. Arguably, the rise of Scala, Clojure, and other languages designed to run on the Java Virtual Machine (JVM) and provide access to Java standard libraries, but otherwise breaking either partly or completely with Java's imperative syntax and its enforcement of an object-oriented paradigm, marks the endurance of the JVM as a platform but the eclipse of Java as a (paradigmatic) language.

7. The difficulties of larger-scale software production are documented by a management-oriented literature stretching back to the 1970s. The canonical text, mentioned previously, is Frederick P. Brooks Jr., *The Mythical Man-Month: Essays on Software Engineering* (1975). See Brooks (1995), along with Britcher (1999), Rosenberg (2008), Weinberg (1998), and, for useful counterpoint, Ullman (2012).

8. The most widely used term in both professional software development and the discipline of computer science is "higher-level language," a spatial metaphor used to describe abstraction from the "lower" level of hardware operation codes. Academic researchers use the historical metaphor of the generation in a similar, if perhaps also more sensible way: a first generation of purely numerically represented instructions ("machine code") is followed by a second generation of mnemonic abbreviations ("assembly language"), followed in turn by compiled, hardware-independent algebraic syntaxes and keywords and phrases in the English language ("programming language" as we use the term today). The classification includes a fourth and a fifth generation, which is beyond my purview here. See Martin (1982).

9. See Lee and Girgensohn (1997).

10. See Ward and Smith (1998), 249: "The World-Wide Web is increasingly in-

fluencing the teaching of Computing Science and associated subjects, and Web-related programming topics are now appearing in many syllabuses. Whilst in this respect there has been much development and discussion of Java as a first programming language with many text books now available, JavaScript has been comparatively ignored. [. . .] We propose here that JavaScript is sufficiently rich in concepts to support the teaching of introductory programming, and that it is especially suitable for Multimedia students." See also Mercuri, Herrmann, and Popyack (1998), 176: "Here we report on a course designed to exploit students' burgeoning interest in the World Wide Web (WWW), where we used HTML and JavaScript to teach programming concepts. These languages allow students at different skill levels to work side by side, learning common abstract ideas while implementing them at different levels of complexity, motivated by the rewarding and exciting interactive environment of the WWW."

11. See Weiss (1999), 35: "As a programming language alone, JavaScript's main appeal has been its simple learning curve, but to more experienced programmers it lacks serious muscle-power. There are sharks in these waters—established, mature programming languages such as Perl can now be embedded into some Web browsers [. . .] For a seasoned developer, the prospect of combining client-side Perl—with its agile handling of advanced programming models—with access to the DOM [Document Object Model] would be lethal to JavaScript. We will likely see the migration of other scripting language into the Web client as well, including Python, TCL, SmallTalk, and perhaps more."

12. On the early history of Netscape, see Reid (1997), Cusumano and Yoffie (1998), Quittner and Slatalla (1998), and Clark (1999).

13. See Hamilton (1999).

14. See Nofre, Priestley, and Alberts (2014).

15. Because it was also created by Ryan Dahl, the original creator of Node.js, specifically to address the latter's shortcomings, Deno has received more attention and adoption than it might have otherwise. However, given the truly massive size of the developer community and ecosystem organized around Node.js, it is unlikely that Deno will replace it anytime soon.

16. See, for example, Landow (1992).

17. See Clemmons (2015) and French-Owen (2016). One cannot say that in 2023, at the time of this writing, there has been any real deceleration in the general pace of development in JavaScript language development, the JavaScript developer community, and the JavaScript developer ecosystem. However, the disruptive introduction of apparently entirely new structural paradigms within the JavaScript ecosystem itself appears to have come to an end. The consistency of the yearly release schedule for new editions of ECMAScript has been a stabilizing factor, as has the steady, dominating evolution of Node.js, which reached version 18 in 2022, and of the AngularJS, Vue.js, and React frameworks, all of which appear to have successfully adapted to the challenge represented by Web-Assembly, a low-level instruction format for a virtual machine that runs in a web

browser, allowing developers to write directly executable web application code in languages other than JavaScript.

18. "JavaScript Affogato" is a variation on (and homage to) the witty phrasing of Reginald Brathwaite in a series of advanced, theoretically sophisticated books on JavaScript. See Braithwaite (2013) and Braithwaite (2015).

Chapter Six

1. See Lewis-Kraus (2016).

2. See Swisher (2018) and Tufekci (2018).

3. See Russell and Vinsel (2018) and Fidler and Russell (2018).

4. Willis (2012) claims that Patrick Debois devised the term "DevOps" in 2009, not in 2008, as Kim, Behr, and Spafford suggest in *The Phoenix Project* (Kim, Behr, and Spafford 2014, 354).

5. See Allspaw and Hammond (2009). I quote here from the PDF slide presentation provided along with the abstract and speaker bios at the Velocity conference website. For a video recording of the presentation, see "10+ Deploys Per Day: Dev and Ops Cooperation at Flickr" (2009).

6. Srnicek and Williams (2016); Bastani (2019); Danaher (2019).

7. Munn (2022), 117. Munn is quoting Schwartz (2019).

8. Munn's source, Schwartz (2019), identifies the six women as "Kathleen McNulty [Kathleen Antonelli], Frances Bilas [Frances Spence], Betty Jean Jennings [Jean Bartik], Ruth Lichterman [Ruth Teitelbaum], Elizabeth Snyder [Betty Holberton], and Marlyn Wescoff [Marlyn Meltzer]." In my own text here, I have used their names as they appear in Kleiman (2022).

BIBLIOGRAPHY

"10+ Deploys per Day: Dev and Ops Cooperation at Flickr." 2009. *YouTube*.

Abbate, Janet. 2000. *Inventing the Internet*. MIT Press.

———. 2012. *Recoding Gender: Women's Changing Participation in Computing*. MIT Press.

Abebe, Surafel Lemma, Sonia Haiduc, Paolo Tonella, and Andrian Marcus. 2011. "The Effect of Lexicon Bad Smells on Concept Location in Source Code." In *2011 IEEE 11th International Working Conference on Source Code Analysis and Manipulation,* Williamsburg, VA, 125–34. IEEE. https://doi.org/10.1109/SCAM.2011.18.

Abrahams, Paul W. 1974. "Improving the Control Structure of SNOBOL4." *ACM SIGPLAN Notices* 9 (5): 10. https://doi.org/10.1145/987413.987415.

Adams, Charles W., and Stanley Gill. 1954. *Digital Computers: Business Applications—Notes from a Special Summer Program*. MIT Press.

Adams, Charles W., and J. H. Laning Jr. 1954. "The MIT Systems of Automatic Coding: Comprehensive, Summer Session, and Algebraic." In *Symposium on Automatic Programming for Digital Computers, Office of Naval Research, Department of the Navy, Washington, D.C., 13–14 May 1954,* 40–68. US Dept. of Commerce, Office of Technical Services.

Alba, Davey. 2016. "The AP Finally Realizes It's 2016, Will Let Us Stop Capitalizing 'Internet'." *Wired*, April. http://www.wired.com/2016/04/ap-finally-realizes-2016-will-let-us-stop-capitalizing-internet/.

Alexiou, Margaret. 1990. "Greek Philology: Diversity and Difference." *Comparative Literature Studies* 27 (1): 53–61. http://www.jstor.org/stable/40246729.

Allspaw, John, and Paul Hammond. 2009. "10+ Deploys per Day: Dev and Ops Cooperation at Flickr." Velocity: Web Performance and Operations Conference, San Jose, CA.

"Amazon Simple Storage Service API Reference." n.d. Amazon Web Services. http://docs.aws.amazon.com/AmazonS3/latest/API.

Ames, Morgan G. 2019. *The Charisma Machine: The Life, Death, and Legacy of One Laptop per Child*. MIT Press.

Andreessen, Marc. 1998. "Innovators of the Net: Brendan Eich and JavaScript." https://web.archive.org/web/20080208124612/http://wp.netscape.com/comp rod/columns/techvision/innovators_be.html.

———. 2011. "Why Software Is Eating the World." *Wall Street Journal*, August.

Aronowitz, Stanley, and William DiFazio. 1994. *The Jobless Future: Sci-Tech and the Dogma of Work*. University of Minnesota Press.

Ashkenas, Jeremy. 2016. "List of Languages That Compile to JS." GitHub. https:// github.com/jashkenas/coffeescript/wiki/list-of-languages-that-compile-to-JS.

"Babel: A Compiler for Writing Next Generation JavaScript." 2016. https://babeljs .io/.

Backus, John W. 1954. "The IBM 701 Speedcoding System." *Journal of the ACM* 1 (1): 4–6. https://doi.org/10.1145/320764.320766.

———. 1959. "The Syntax and Semantics of the Proposed International Algebraic Language of the Zurich ACM-GAMM Conference." In *Proceedings of the International Conference on Information Processing*, 125–32. UNESCO.

———. 1981. "The History of Fortran I, II and III." In *History of Programming Languages*, edited by Richard L. Wexelblat, 25–74. New York: Academic Press.

Backus, John W., and Harlan Herrick. 1954. "IBM 701 Speedcoding and Other Automatic-Programming Systems." In *Symposium on Automatic Programming for Digital Computers, Office of Naval Research, Department of the Navy, Washington, D.C., 13–14 May 1954*, 106–13. US Dept. of Commerce, Office of Technical Services.

Barnett, Michael P. 1970. "SNAP: A Programming Language for Humanists." *Computers and the Humanities* 4 (4): 225–40. https://www.jstor.org/stable/ i30199360.

Bastani, Aaron. 2019. *Fully Automated Luxury Communism: A Manifesto*. Verso.

Bellucci, Sergio. 2005. *E-Work: Lavoro, Rete, Innovazione*. Rome: Derive Approdi.

Bemer, R. W. 1969. "A Politico-Social History of Algol (With a Chronology in the Form of a Log Book)." In *Annual Review of Automatic Programming 5*, 151–237. Pergamon.

Benanav, Aaron. 2020. *Automation and the Future of Work*. New York: Verso.

Benjamin, Ruha. 2019. *Race after Technology: Abolitionist Tools for the New Jim Code*. Polity.

Bentley, Jon, Don Knuth, and Doug McIlroy. 1986. "Programming Pearls." *Communications of the ACM* 29 (6): 471–83. https://doi.org/10.1145/5948.315654.

Bergin, Thomas J. 2007. "A History of the History of Programming Languages." *Communications of the ACM* 50 (5): 69–74. https://doi.org/10.1145/1230819.12 30841.

Bergin, Thomas J., and Richard G. Gibson, eds. 1996. *History of Programming Languages II*. ACM Press, Addison-Wesley.

Berkeley, Edmund C., and Daniel G. Bobrow, eds. 1964. *The Programming Language LISP: Its Operation and Applications*. Information International.

Berry, David M. 2011. *The Philosophy of Software: Code and Mediation in the Digital Age*. Palgrave Macmillan.

Bishop, Bryan. 2012. "Discovering the Hidden Ascii Art in the Pages of the Web." *The Verge*, April. http://www.theverge.com/2012/4/25/2976042/discovering -hidden-ascii-art-in-the-pages-of-the-web.

Bix, Amy Sue. 2002. *Inventing Ourselves Out of Jobs? America's Debate over Technological Unemployment, 1929–1981*. Johns Hopkins University Press.

Blackmore, David S. T. 2009. "Fubar." *The Seafaring Dictionary: Terms, Idioms and Legends of the Past and Present*. McFarland.

Boden, Chris. 2016. "What Is the Funniest Comment You've Ever Seen in Source Code?" *Reddit*. https://www.reddit.com/r/linux/comments/413enq/what_is_ the_funniest_comment_youve_ever_seen_in/.

Böhm, Corrado, and Giuseppe Jacopini. 1966. "Flow Diagrams, Turing Machines and Languages with Only Two Formation Rules." *Communications of the ACM* 9 (5): 366–71. https://doi.org/10.1145/355592.365646.

Bové, Paul A. 2009. "Philology and Poetry: The Case against Descartes." *Law and Literature* 21 (2): 149–68. http://www.jstor.org/stable/10.1525/lal.2009.21 .2.149.

Bowles, Nellie. 2019. "The Strange Experience of Being Australia's First Tech Billionaires." *New York Times*, February.

Braithwaite, Reginald. 2013. *JavaScript Allongé*. Leanpub.

——. 2015. *JavaScript Spessore*. Leanpub.

Braman, Sandra. 2011. "The Framing Years: Policy Fundamentals in the Internet Design Process, 1969–1979." *Information Society* 27 (5): 295–310. https://doi .org/10.1080/01972243.2011.607027.

Bratton, Benjamin H. 2015. *The Stack: On Software and Sovereignty*. MIT Press.

Brenner, Robert. 2006. *The Economics of Global Turbulence: The Advanced Capitalist Economies from Long Boom to Long Downturn, 1945–2005*. Verso.

Britcher, Robert N. 1999. *The Limits of Software: People, Projects, and Perspectives*. Addison-Wesley.

Bron, Dan. 2017. "What Is the History of the Term 'Metasyntactic Variable'?" *Stack Exchange: English Language and Usage*. http://english.stackexchange .com/questions/375614/what-is-the-history-of-the-term-metasyntactic-varia ble/375630.

Bronson, Po. 1998. *The First $20 Million Is Always the Hardest*. Avon Books.

Brooks, Frederick P., Jr. 1995. *The Mythical Man-Month: Essays on Software Engineering*. 20th Anniversary Edition. Addison-Wesley.

Brown, J. H., and John W. Carr. 1954. "Automatic Programming and Its Development on the MIDAC." In *Symposium on Automatic Programming for Digital Computers, Office of Naval Research, Department of the Navy, Washington, D.C., 13–14 May 1954*, 84–97. US Dept. of Commerce, Office of Technical Services.

Burns, Ken, and Lynn Novick. 2007. "FUBAR." *The War*. Public Broadcasting Service.

Burris, Beverly H. 1993. *Technocracy at Work*. State University of New York Press.

Bush, Vannevar. 1945. "As We May Think." *Atlantic Monthly*, July. http://www.theatlantic.com/magazine/archive/1945/07/as-we-may-think/303881/.

Campbell-Kelly, Martin. 2004. *From Airline Reservations to Sonic the Hedgehog: A History of the Software Industry*. MIT Press.

Carpenter, Brian E., and Craig Partridge. 2010. "Internet Requests for Comments (RFCs) as Scholarly Publications." *ACM SIGCOMM Computer Communication Review* 40 (1): 31. https://doi.org/10.1145/1672308.1672315.

Carson, Biz. 2016. "Silicon Valley Was Worried about the Wrong Bubble." *Business Insider*, November. https://www.businessinsider.com/how-silicon-valley-missed-trump-2016-11.

Cassel, David. 2016. "Evolve or Die: Java, C++ Confront Newcomers on the TIOBE Index." *The New Stack*. https://thenewstack.io/evolve-die-popular-programming-languages-confront-newcomers-tiobe-index/.

———. 2017. "COBOL Is Everywhere. Who Will Maintain It?" *The New Stack*, May. https://thenewstack.io/cobol-everywhere-will-maintain/.

Ceruzzi, Paul E. 2003. *A History of Modern Computing*. 2nd ed. MIT Press.

Chapman, Robert L., Barbara Ann Kipfer, and Harold Wentworth. 1995. *Dictionary of American Slang*. 3rd ed. New York: HarperCollins.

Chapman, Robert L., and Harold Wentworth. 1986. *New Dictionary of American Slang*. Harper & Row.

Chartier, Rob. 2004. "Longest & Shortest Types in .NET." *Contemplation*. https://weblogs.asp.net/rchartier/68350.

Chun, Wendy Hui Kyong. 2013. *Programmed Visions: Software and Memory*. MIT Press.

Clark, Jim. 1999. *Netscape Time: The Making of the Billion-Dollar Start-up That Took on Microsoft*. St. Martin's Press.

Clarke, Roger. 2012. "Systems Administrators [Are] History." http://www.rogerclarke.com/II/SAH-12.html.

Clemmons, Eric. 2015. "JavaScript Fatigue." *Medium*. https://medium.com/@ericclemmons/JavaScript-fatigue-48d4011b6fc4#.c0ve3n241.

"COBOL Blues." n.d. *Reuters Graphics*. http://fingfx.thomsonreuters.com/gfx/rngs/USA-BANKS-COBOL/010040KH18J/index.html.

"COBOL: Initial Specifications for a Common Business Oriented Language." 1960. US Dept. of Defense. http://bitsavers.informatik.uni-stuttgart.de/pdf/codasyl/COBOL_Report_Apr60.pdf.

Coffey, Lauren. 2023. "Southern New Hampshire Shuttering Kenzie Academy Boot Camp." *Inside Higher Ed*, August.

Collins, Keith. 2016. "The Code That Took America to the Moon Was Just Published to Github, and It's Like a 1960s Time Capsule." *Quartz*, July. http://qz .com/726338/the-code-that-took-america-to-the-moon-was-just-published-to -github-and-its-like-a-1960s-time-capsule/.

COMIT Programmers' Reference Manual. 1962. 2nd corrected ed. MIT Research Laboratory of Electronics, Computation Center.

"Computing Conversations with Brendan Eich." 2012. YouTube. https://www.you tube.com/watch?v=IPxQ9kEaF8c.

Copeland, Jules. 2010. "Stop Using Foo Bar!" *Ruby Forum: Ruby on Rails*. https:// www.ruby-forum.com/topic/207096.

Cormen, Thomas H., Charles E. Leiserson, Ronald L. Rivest, and Clifford Stein. 2009. *Introduction to Algorithms*. 3rd ed. MIT Press.

Corré, Alan D. 2010. *Icon Programming for Humanists*. 2nd ed. Goal-Directed Press. http://www2.cs.uidaho.edu/~jeffery/icon/humanists/humanist.pdf.

Cortada, James W. 2004. *The Digital Hand: How Computers Changed the Work of American Manufacturing, Transportation, and Retail Industries*. Oxford University Press.

———. 2006. *The Digital Hand, Volume II: How Computers Changed the Work of American Financial, Telecommunications, Media, and Entertainment Industries*. Oxford University Press.

———. 2008. *The Digital Hand, Volume III: How Computers Changed the Work of American Public Sector Industries*. Oxford University Press.

Coupland, Douglas. 2008. *Microserfs*. Harper Perennial.

Cox, Geoff, and Alex McLean. 2013. *Speaking Code: Coding as Aesthetic and Political Expression*. MIT Press.

Crockford, Douglas. 2003. "The Little JavaScripter." *Douglas Crockford's World Wide Web*. http://www.crockford.com/JavaScript/little.html.

———. 2008. *JavaScript: The Good Parts*. O'Reilly.

Croll, Angus. 2014. "A Book Nerd's Guide to JavaScript." dotJS 2014, Paris. https: //www.youtube.com/watch?v=uub1wARuSHQ.

———. 2015. *If Hemingway Wrote JavaScript*. No Starch Press.

Cusumano, Michael A., and David B. Yoffie. 1998. *Competing on Internet Time: Lessons from Netscape and Its Battle with Microsoft*. Free Press.

Da, Nan Z. 2019. "The Computational Case against Computational Literary Studies." *Critical Inquiry* 45 (3): 601–39.

Danaher, John. 2019. *Automation and Utopia: Human Flourishing in a World without Work*. Harvard University Press.

Darlington, J. L. 1965. "Machine Methods for Proving Logical Arguments Expressed in English." *Mechanical Translation* 8 (3–4): 41–67.

Davis, Martin, and Hilary Putnam. 1960. "A Computing Procedure for Quantification Theory." *Journal of the ACM* 7 (3): 201–15. https://doi.org/10.1145/3210 33.321034.

Del Rey, Jason. 2023. "Inside the Battle for the Future of Amazon." *Vox*, January. https://www.vox.com/recode/2023/1/19/23562702/amazon-layoffs-andy -jassy-day-2-alexa.

Desautels, E. J., and D. K. Smith. 1967. "An Introduction to the String Manipulation Language SNOBOL." In *Programming Systems and Languages*, edited by S. Rosen, 419–54. McGraw-Hill.

"Did COBOL Have 250 Billion Lines of Code and 1 Million Programmers, as Late as 2009?" 2017. *Stack Exchange.* https://skeptics.stackexchange.com /questions/5114/did-cobol-have-250-billion-lines-of-code-and-1-million -programmers-as-late-as-2.

Dijkstra, Edsger W. 1968. "Go-to Statement Considered Harmful." *Communications of the ACM* 11 (3): 147–48. https://doi.org/10.1145/362929.362947.

——. 1982. "EWD 498: How Do We Tell Truths That Might Hurt?" In *Selected Writings on Computing: A Personal Perspective*, 129–31. Springer-Verlag.

"Drupal API." n.d. *Drupal API.* https://api.drupal.org.

Druseikis, Frederick C., and John N. Doyle. 1974. "A Procedural Approach to Pattern Matching in Snobol4." In *ACM 74: Proceedings of the 1974 Annual Conference*, 311–17. ACM Press. https://doi.org/10.1145/800182.810418.

Dunn, Richard. 1973. "SNOBOL4 as a Language for Bootstrapping a Compiler." *ACM SIGPLAN Notices* 8 (5): 28. https://doi.org/10.1145/986948.986951.

Eastlake, D., C. Manros, and E. Raymond. 2001. "Etymology of 'Foo'." Request for {Comments} ({RFC}) 3092. Network Working Group. https://www.ietf.org /rfc/rfc3092.txt.

"ECMAScript: A General Purpose, Cross-Platform Programming Language. Standard Ecma-262, June 1997." 1997. ECMA. http://www.ecma-international .org/publications/files/ECMA-ST-ARCH/ECMA-262,%201st%20edition,%20 June%201997.pdf.

"ECMAScript Language Specification: Standard ECMA-262, 3rd Edition." 1999. ECMA. http://www.ecma-international.org/publications/files/ECMA-ST -ARCH/ECMA-262,%203rd%20edition,%20December%201999.pdf.

"ECMAScript Language Specification: Standard ECMA-262, 5th Edition." 2009. ECMA. http://www.ecma-international.org/publications/files/ECMA-ST-AR CH/ECMA-262%205th%20edition%20December%202009.pdf.

"ECMAScript Language Specification: Standard ECMA-262, 6th Edition." 2015. ECMA. http://www.ecma-international.org/ecma-262/6.0/.

Eco, Umberto. 1995. *The Search for the Perfect Language.* Translated by James Fentress. Blackwell.

Elkin, Frederick. 1946. "The Soldier's Language." *American Journal of Sociology* 51 (5): 414–22. http://www.jstor.org/stable/2771105.

Emmer, Mark B. n.d. "Preface to SNOBOL4: The SNOBOL4 Language for the 8086/8088 Computer Family." Catspaw Inc.

Emmer, Mark B. 1984. "Implementing SNOBOL4 for the 8086 Micro-Computer Family." *ACM SIGSMALL Newsletter* 10 (4): 12–20. https://doi.org/10.1145/11 64662.1164663.

Ensmenger, Nathan. 2010. *The Computer Boys Take Over: Computers, Programmers, and the Politics of Technical Expertise.* MIT Press.

Etzkorn, Letha H., Carl G. Davis, and Lisa L. Bowen. 2001. "The Language of Comments in Computer Software: A Sublanguage of English." *Journal of Pragmatics* 33 (11): 1731–56. https://doi.org/10.1016/S0378-2166(00)00068-0.

Eve, Martin Paul, and Joe Street. 2018. "The Silicon Valley Novel." *Literature & History* 27 (1): 81–97. https://doi.org/10.1177/0306197318755680.

Evens, Aden. 2015. *Logic of the Digital.* Bloomsbury Academic.

Everett, John. 1996. "Foobar." *Alt.folklore.computers.* https://groups.google.com/d/msg/alt.folklore.computers/8tzUJHuGamI/R7wpsm6T7rMJ.

Farber, D. J., R. E. Griswold, and I. P. Polonsky. 1964. "SNOBOL, A String Manipulation Language." *Journal of the ACM* 11 (1): 21–30. https://doi.org/10.1145/321203.321207.

Felleisen, Matthias, Robert Bruce Findler, Matthew Flatt, and Shriram Krishnamurthi. 2018. *How to Design Programs: An Introduction to Programming and Computing.* 2nd ed. MIT Press.

Fidler, Bradley, and Andrew L. Russell. 2018. "Financial and Administrative Infrastructure for the Early Internet: Network Maintenance at the Defense Information Systems Agency." *Technology and Culture* 59 (4): 899–924. https://doi.org/10.1353/tech.2018.0090.

Flanagan, David. 2011. *JavaScript: The Definitive Guide.* 6th ed. O'Reilly.

" 'Foo', 'Bar'—Does This Actually Help Anyone—Ever?" 2014. *Reddit.* https://redd.it/2jelzu.

Ford, Paul. 2015. "What Is Code?" *Bloomberg Businessweek*, June. https://www.bloomberg.com/graphics/2015-paul-ford-what-is-code/.

Forte, Allen. 1967. "The Programming Language SNOBOL3: An Introduction." *Computers and the Humanities* 1 (5): 157–63. http://www.jstor.org/stable/30199238.

———. 1968. *SNOBOL3 Primer: An Introduction to the Computer Programming Language.* 2nd printing. MIT Press.

The Fortran Automatic Coding System for the IBM 704: Programmer's Reference Manual. 1956. New York: International Business Machines Corporation. http://www.fortran.com/FortranForTheIBM704.pdf.

Frabetti, Federica. 2015a. *Software Theory: A Cultural and Philosophical Study.* Rowman & Littlefield International.

———. 2015b. *Software Theory: A Cultural and Philosophical Study.* Rowman & Littlefield International.

Franklin, Seb. 2015. *Control: Digitality as Cultural Logic.* MIT Press.

French-Owen, Calvin. 2016. "The Deep Roots of JavaScript Fatigue." *Segment Blog.* https://segment.com/blog/the-deep-roots-of-js-fatigue/.

Friedman, Daniel P., and Matthias Felleisen. 1996. *The Little Schemer.* 4th ed. MIT Press.

"Fubar." 2002. *Oxford Essential Dictionary of the U.S. Military.* Oxford University Press. http://www.oxfordreference.com/view/10.1093/acref/9780199891580.001.0001/acref-9780199891580.

"Fubar, Adj." 2016. *Oxford English Dictionary Online*. Oxford University Press.

Fuchs, Christian. 2014. *Digital Labour and Karl Marx*. Routledge.

———. 2016. *Reading Marx in the Information Age: A Media and Communication Studies Perspective on Capital, Volume 1*. Routledge, Taylor & Francis Group.

Fuchs, Christian, and Vincent Mosco, eds. 2016. *Marx in the Age of Digital Capitalism*. Haymarket Books.

"Funny Things Seen in Source Code and Documentation." 2014. *WikiWikiWeb*. http://wiki.c2.com/?FunnyThingsSeenInSourceCodeAndDocumentation.

Gabriel, Richard P, and Ron Goldman. 2000. "Mob Software: The Erotic Life of Code." ACM Conference on Object-Oriented Programming, Systems, Languages, and Applications, Minneapolis. http://www.dreamsongs.com/Mob Software.html.

Galley, S. W., and Greg Pfister. 1979. *The MDL Programming Language*. Laboratory for Computer Science, MIT.

Garry, Chris. 2009. "Apollo-11." https://github.com/chrislgarry/Apollo-11.

Genette, Gérard. 1987. *Seuils*. Éditions du Seuil.

———. 1997. *Paratexts: Thresholds of Interpretation*. Translated by J. E. Lewin. Cambridge University Press.

Gill, Stanley. 1954. "General Discussion at End of Thursday Afternoon Session." In *Symposium on Automatic Programming for Digital Computers, Office of Naval Research, Department of the Navy, Washington, D.C., 13–14 May 1954*, 98. US Dept. of Commerce, Office of Technical Services.

Gimpel, James F. 1972. "Blocks—A New Datatype for Snobol4." *Communications of the ACM* 15 (6): 438–47. https://doi.org/10.1145/361405.361410.

"GitHub Developer: REST API V3." n.d. *GitHub Developer*. https://developer.git hub.com/v3/.

Goldberg, Adele. 1987. "Programmer as Reader." *IEEE Software* 4 (5): 62–70. https://doi.org/10.1109/MS.1987.231775.

Goldratt, Eliyahu M., and Jeff Cox. 2014. *The Goal: A Process of Ongoing Improvement*. 30th Anniversary Edition. North River Press.

Golumbia, David. 2009. *The Cultural Logic of Computation*. Harvard University Press.

———. 2016. *The Politics of Bitcoin: Software as Right-Wing Extremism*. University of Minnesota Press.

Goodliffe, Pete. 2007. *Code Craft: The Practice of Writing Excellent Code*. No Starch Press.

Gorn, Saul. 1954. "Planning Universal Semi-Automatic Coding." In *Symposium on Automatic Programming for Digital Computers, Office of Naval Research, Department of the Navy, Washington, D.C., 13–14 May 1954*, 74–83. US Dept. of Commerce, Office of Technical Services.

———. 1960. "Introductory Speech—Rapport Général." In *Information Processing: Proceedings of the International Conference on Information Processing, UNESCO, Paris 15–20 June 1959*, 117–18. UNESCO, R. Oldenbourg, Butterworths.

Gosling, James, Bill Joy, Guy Steele, Gilad Bracha, Alex Buckley, and Daniel Smith. 2019. "The Java® Language Specification: Java SE 12 Edition." Oracle Corp.

Grafton, Anthony. 1999. *The Footnote: A Curious History*. Harvard University Press.

Greenbaum, Joan M. 1979. *In the Name of Efficiency: Management Theory and Shopfloor Practice in Data-Processing Work*. Temple University Press.

Grier, David Alan. 2007. *When Computers Were Human*. Princeton University Press.

Griswold, R. E. 1962. "The Separation of Flow Graphs." Bell Laboratories Unpublished Technical Memorandum 62-3344-3.

———. 1974. "Suggested Revisions and Additions to the Syntax and Control Mechanisms of Snobol4." *ACM SIGPLAN Notices* 9 (2): 7. https://doi.org/10.1145/987424.987426.

———. 1981. "SNOBOL Session." In *History of Programming Languages*, edited by Richard L. Wexelblat, 601–60. Academic Press.

———. 1985. "SNOBOL: A Personal Perspective." In *ICEBOL 85 Proceedings: The Proceedings of the 1985 International Conference on English Language and Literature Applications of SNOBOL and SPITBOL*, 1–5. Dakota State College.

Griswold, R. E., J. F. Poage, and I. P. Polonsky. 1971. *The SNOBOL4 Programming Language*. 2nd ed. Prentice-Hall.

Griswold, R. E., and I. P. Polonsky. 1962. "The Classification of the States of a Finite Markov Chain." Holmdel, NJ. Bell Laboratories Unpublished Technical Memorandum 62-3344-3.

Grouchnikov, Kiril. 2007. "And the Longest JRE Class Name Is . . ." *Pushing Pixels*. http://www.pushing-pixels.org/2007/11/07/and-the-longest-jre-class-name-is.html.

Haigh, Thomas. 2014. "We Have Never Been Digital." *Communications of the ACM* 57 (9): 24–28. https://doi.org/10.1145/2644148.

Hamilton, Walter. 1999. "Hooked on Speed: How Day Trading Works." *Los Angeles Times*, February. http://articles.latimes.com/1999/feb/21/business/fi-10174.

Hart, Timothy P, and Michael I. Levin. 1964. "LISP Exercises." http://hdl.handle.net/1721.1/5924.

Heller, Jack, and George W. Logemann. 1966. "PL/I: A Programming Language for Humanities Research." *Computers and the Humanities* 1 (2): 19–27. https://www.jstor.org/stable/30199201.

Hewitt, Carl, Peter Bishop, and Richard Steiger. 1973. "A Universal Modular ACTOR Formalism for Artificial Intelligence." In *Proceedings of the Third International Joint Conference on Artificial Intelligence*, 235–45. Stanford, CA.

Hicks, Mar. 2017. *Programmed Inequality: How Britain Discarded Women Technologists and Lost Its Edge in Computing*. MIT Press.

Hockey, Susan M. 1985. *Snobol Programming for the Humanities*. Clarendon Press of Oxford University Press.

Holden, Richard. 2010. "Digital." Oxford {English} {Dictionary}. *Word Stories.* http://public.oed.com/aspects-of-english/word-stories/digital/.

Holen, Vidar. 2004. "Wordcount: Occurrences of Words in the Linux Kernel Source Code over Time." http://www.vidarholen.net/contents/wordcount/.

Holman, Bill. 1938. *Smokey Stover and the Fire Chief of Foo.* Whitman Publishing.

Holquist, Michael. 2011. "The Place of Philology in an Age of World Literature." *Neohelicon* 38 (2): 267–87. https://doi.org/10.1007/s11059-011-0096-7.

Hopper, Grace M. 1954. "Automatic Programming—Definitions." In *Symposium on Automatic Programming for Digital Computers, Office of Naval Research, Department of the Navy, Washington, D.C., 13–14 May 1954*, 1–5. US Dept. of Commerce, Office of Technical Services.

———. 1956. "The Interlude 1954–1956." In *Symposium on Advanced Programming Methods for Digital Computers: Washington, D.C., June 28, 29, 1956*, 1–2. Office of Naval Research, Dept. of the Navy.

Hui, Yuk. 2016. *On the Existence of Digital Objects.* University of Minnesota Press.

"IBM Speedcoding System (from IBM Manual)." 1983. *IEEE Annals of the History of Computing* 5 (2): 135–39. https://doi.org/10.1109/MAHC.1983.10071.

An Introduction to COMIT Programming. 1961. MIT Research Laboratory of Electronics, Computation Center.

Isaac, Earl J. 1952. "Machine Aids to Coding." In *Proceedings of the 1952 ACM National Meeting (Toronto)*, 17–18. ACM '52. ACM. https://doi.org/10.1145/80 0259.808983.

"ISO/IEC 9899:2018 Information Technology—Programming Languages—C." 2018. International Organization for Standardization.

"ISO/IEC 14882:2017 Programming Languages—C++." 2017. International Organization for Standardization.

Jeffery, Clinton, Phillip Thomas, Sudarshan Gaikaiwari, and John Goettsche. 2016. "Integrating Regular Expressions and SNOBOL Patterns into String Scanning: A Unifying Approach." In *SAC '16 Proceedings of the 31st Annual ACM Symposium on Applied Computing*, 1974–79. ACM Press. https://doi.org/10.1145/2851613.2851752.

Joque, Justin. 2016. "The Invention of the Object: Object Orientation and the Philosophical Development of Programming Languages." *Philosophy & Technology* 29 (4): 335–56. https://doi.org/10.1007/s13347-016-0223-5.

Kernighan, Brian W. 1974. *The Elements of Programming Style.* 1st ed. McGraw-Hill.

Kernighan, Brian W., and P. J. Plauger. 1978. *The Elements of Programming Style.* 2nd ed. New York: McGraw-Hill.

Kim, Gene, Kevin Behr, and George Spafford. 2014. *The Phoenix Project: A Novel about IT, DevOps, and Helping Your Business Win.* Rev. ed. with new resource guide. IT Revolution Press.

Kleiman, Kathy. 2022. *Proving Ground: The Untold Story of the Six Women Who Programmed the World's First Modern Computer.* Grand Central Publishing.

Knuth, Donald E. 1964. "Backus Normal Form vs. Backus Naur Form." *Communications of the ACM* 7 (12): 735–36. https://doi.org/10.1145/355588.365140.

——. 1974. "Computer Programming as an Art." *Communications of the ACM* 17 (12): 667–73. https://doi.org/10.1145/361604.361612.

——. 1984. "Literate Programming." *Computer Journal* 27 (2): 97–111. https://doi.org/10.1093/comjnl/27.2.97.

——. 1992a. "Computer Programming as an Art." In *Literate Programming*, 1–16. Center for the Study of Language and Information.

——. 1992b. *Literate Programming*. Center for the Study of Language and Information.

——. 1996. "Computer Programming and Computer Science." In *Selected Papers on Computer Science*, 2–3. CSLI Publications.

——. 1997. *The Art of Computer Programming, Vol. 1: Fundamental Algorithms*. 3rd ed. Addison-Wesley.

——. 2003. "The Early Development of Programming Languages." In *Selected Papers on Computer Languages*. CSLI Publications.

Knuth, Donald E., and Jack N. Merner. 1961. "Algol 60 Confidential." *Communications of the ACM* 4 (6): 268–72. https://doi.org/10.1145/366573.366599.

Knuth, Donald E., and Luis Trabb Pardo. 1976. "The Early Development of Programming Languages." Stanford, CA.

Ko, Amy J. 2016. "What Is a Programming Language, Really?" In *PLATEAU 2016: Proceedings of the 7th International Workshop on Evaluation and Usability of Programming Languages and Tools*, 32–33. ACM Press. https://doi.org/10.1145/3001878.3001880.

Kraft, Philip. 1977. *Programmers and Managers: The Routinization of Computer Programming in the United States*. Springer-Verlag.

Kurtz, Thomas E. 1981. "BASIC Session." In *History of Programming Languages*, edited by Richard L. Wexelblat, 515–50. Academic Press.

Landow, George P. 1992. *Hypertext: The Convergence of Contemporary Critical Theory and Technology*. Johns Hopkins University Press.

——. 2006. *Hypertext 3.0: Critical Theory and New Media in an Era of Globalization*. 3rd ed. Johns Hopkins University Press.

Laning, J. H., Jr., and N. Zierler. 1954. "A Program for Translation of Mathematical Equations for Whirlwind I." MIT. http://archive.computerhistory.org/resources/text/Fortran/102653982.05.01.acc.pdf.

Latour, Bruno. 2004. "Why Has Critique Run Out of Steam? From Matters of Fact to Matters of Concern." *Critical Inquiry* 30 (2): 225–48. https://doi.org/10.1086/421123.

Lee, Alison, and Andreas Girgensohn. 1997. "Developing Collaborative Applications Using the World Wide Web 'Shell'." In *CHI EA '97: CHI '97 Extended Abstracts on Human Factors in Computing Systems*, 144. ACM Press. https://doi.org/10.1145/1120212.1120314.

Lee, C. Y. 1961. "An Algorithm for Path Connections and Its Applications." *IEEE*

Transactions on Electronic Computers EC-10 (3): 346–65. https://doi.org/10
.1109/TEC.1961.5219222.

Lennon, Brian. 2013. "New Stationary States: Real Time and History's Disquiet."
Symplokē 21 (1–2): 179–93. http://www.jstor.org/stable/10.5250/symploke.21
.1-2.0179.

——. 2014. "Machine Translation: A Tale of Two Cultures." In *A Companion to
Translation Studies*, edited by Sandra Bermann and Catherine Porter, 133–
46. John Wiley & Sons. http://doi.wiley.com/10.1002/9781118613504.ch10.

Levy-Eichel, Mordechai, and Daniel Scheinerman. 2022. "Digital Humanists
Need to Learn How to Count." *Chronicle of Higher Education*, May.

Lewis-Kraus, Gideon. 2016. "Bubble Indemnity." *New York Times Magazine*, May.
http://www.nytimes.com/2016/05/15/magazine/bubble-indemnity.html.

Light, Jennifer S. 1999. "When Computers Were Women." *Technology and Culture* 40 (3): 455–83. http://muse.jhu.edu/journals/technology_and_culture/
v040/40.3light.html.

Lindsey, C. H. 1968. "ALGOL 68 with Fewer Tears." *ALGOL Bulletin* 28 (July):
9–49. http://dl.acm.org/citation.cfm?id=1061116.

——. 1972. "ALGOL 68 with Fewer Tears." *Computer Journal* 15 (2): 176–88. https:
//doi.org/10.1093/comjnl/15.2.176.

Lopes, Cristina Videira. 2014. *Exercises in Programming Style*. CRC Press, Taylor
& Francis Group.

——. 2021. *Exercises in Programming Style*. 2nd ed. CRC Press, Taylor & Francis
Group.

Mackenzie, Adrian. 2006. *Cutting Code: Software and Sociality*. Peter Lang.

Marino, Mark C. 2006. "Critical Code Studies." *Electronic Book Review*, December. http://www.electronicbookreview.com/thread/electropoetics/codology.

——. 2020. *Critical Code Studies*. MIT Press.

Markov, A. A., and N. M. Nagorny. 1988. *The Theory of Algorithms*. Translated by
M. Greendlinger. Kluwer Academic.

Martin, James. 1982. *Application Development without Programmers*. Prentice-
Hall.

Mauchly, John W. 1988. "Suggested Form for 'BINAC BRIEF CODE'." *Annals of
the History of Computing* 10 (1): 7–18. https://doi.org/10.1109/MAHC.1988.10
004.

Maurer, W. Douglas. 1976. *The Programmer's Introduction to SNOBOL*. Elsevier.

McCarthy, John, Paul W. Abrahams, Daniel J. Edwards, Timothy P. Hart, and
Michael I. Levin. 1962. *Lisp 1.5 Programmer's Manual*. MIT Press. http://www
.softwarepreservation.org/projects/LISP/book/LISP%201.5%20Program
mers%20Manual.pdf.

McCarthy, John, R. Brayton, D. Edwards, P. Fox, L. Hodes, D. Luckham, K. Maling,
D. Park, and S. Russell. 1960. *LISP I Programmer's Manual*. Computation
Center, Research Laboratory of Electronics, MIT. http://www.softwarepreser
vation.org/projects/LISP/book/LISP%20I%20Programmers%20Manual.pdf.

McConnell, Steve. 1993. *Code Complete: A Practical Handbook of Software Construction.* Microsoft Press.

McGee, Russell C. 1957. "Omnicode—A Common Language Programming System." In *Automatic Coding: Proceedings of the Symposium on Automatic Coding, January 24-25, Franklin Institute, Philadelphia,* 57–70. Franklin Institute.

McIlwain, Charlton D. 2020. *Black Software: The Internet and Racial Justice, from the Afronet to Black Lives Matter.* Oxford University Press.

McKinley, Dan. 2009. "From the Annals of Dubious Achievement." *Dan McKinley.* http://mcfunley.com/from-the-annals-of-dubious-achievement.

McMillan, Robert. 2013. "Is Java Losing Its Mojo?" *Wired*, January. http://www.wired.com/2013/01/java-no-longer-a-favorite/.

McMurtrie, Beth. 2022. "AI and the Future of Undergraduate Writing." *Chronicle of Higher Education*, December.

McPeak, Jeremy, and Paul Wilton. 2015. *Beginning JavaScript, 5th Edition.* John Wiley & Sons.

"MediaWiki API:Main Page." n.d. *MediaWiki.* https://www.mediawiki.org/wiki/API:Main_page.

Merchant, Brian. 2018. "The Coders Programming Themselves Out of a Job." *The Atlantic.* https://www.theatlantic.com/technology/archive/2018/10/agents-of-automation/568795/.

Mercuri, Rebecca, Nira Herrmann, and Jeffrey Popyack. 1998. "Using HTML and JavaScript in Introductory Programming Courses." *ACM SIGCSE Bulletin* 30 (1): 176–80. https://doi.org/10.1145/274790.273754.

Metz, Cade. 2011. "The Chef, the Puppet, and the Sexy IT Admin." *Wired*, October.

———. 2016. "Here's How Google Makes Sure It (Almost) Never Goes Down." *Wired*, April.

Montfort, Nick, Patsy Baudoin, John Bell, Ian Bogost, Jeremy Douglass, Mark C. Marino, Michael Mateas, Casey Reas, Mark Sample, and Noah Vawter. 2013. *10 PRINT CHR$(205.5+RND(1));:GOTO 10.* MIT Press.

Montler, Timothy. 1985. "Northwest American Indian Language Data Processing with SNOBOL." In *ICEBOL 85 Proceedings: The Proceedings of the 1985 International Conference on English Language and Literature Applications of SNOBOL and SPITBOL,* 168–96. Dakota State College.

Mother of Five, Indignant. 1937. "Letter to the Editor." *The Tech*, September, 2. http://tech.mit.edu/V57/PDF/V57-N30.pdf.

Mufti, Aamir R. 2010. "Orientalism and the Institution of World Literatures." *Critical Inquiry* 36 (3): 458–93. http://www.jstor.org/stable/10.1086/653408.

Munn, Luke. 2022. *Automation Is a Myth.* Stanford University Press.

Nelson, T. H. 1965a. "A File Structure for the Complex, the Changing, and the Indeterminate." In *ACM '65: Proceedings of the 1965 20th National Conference,* 84–100.

———. 1965b. "Complex Information Processing: A File Structure for the Complex, the Changing and the Indeterminate." In *ACM '65: Proceedings of the 1965 20th National Conference,* 84–100. ACM Press. https://doi.org/10.1145/800197 .806036.

"Netscape and Sun Announce JavaScript, the Open, Cross-Platform Object Scripting Language for Enterprise Networks and the Internet." 1995. https: //web.archive.org/web/20070916144913/http://wp.netscape.com/newsref/pr /newsrelease67.html.

Newsted, Peter R. 1975. *SNOBOL: An Introduction to Programming.* Hayden Book Co.

Nguyen, Clinton. 2015. "What Is 'What Is Code?'" *Vice,* June. https://www.vice .com/en/article/qkv9vd/what-is-what-is-code.

Nicholls, John E. 1975. *The Structure and Design of Programming Languages.* Addison-Wesley.

Noble, Safiya Umoja. 2018. *Algorithms of Oppression: How Search Engines Reinforce Racism.* New York University Press.

Nofre, David, Mark Priestley, and Gerard Alberts. 2014. "When Technology Became Language: The Origins of the Linguistic Conception of Computer Programming, 1950–1960." *Technology and Culture* 55 (1): 40–75. https://doi .org/10.1353/tech.2014.0031.

Olsen, Mark. 1987. "Beyond SNOBOL: The Icon Programming Language." *Computers and the Humanities* 21 (1): 61–66. https://doi.org/10.1007/BF00125225.

"On Foo-Ism." 1938. *The Tech,* January, 2. http://tech.mit.edu/V57/PDF/V57-N57 .pdf.

Ousterhout, J. K. 1998. "Scripting: Higher Level Programming for the 21st Century." *Computer* 31 (3): 23–30. https://doi.org/10.1109/2.660187.

Paine, Jocelyn. 2010a. "More Technonecrophilia with Snobol One-Liners." *Dr. Dobb's Blog.* http://www.drdobbs.com/architecture-and-design/more-techno necrophilia-with-snobol-one-l/228700458.

———. 2010b. "Programs That Transform Their Own Source Code; or: The Snobol Foot Joke." *Dr. Dobb's Blog.* http://www.drdobbs.com/architecture-and-design /programs-that-transform-their-own-source/228701469.

Paulson, William. 2001. "For a Cosmopolitical Philology: Lessons from Science Studies." *SubStance* 30 (3): 101–19. https://doi.org/10.1353/sub.2001.0033.

Perlis, A. J., and H. Samelson. 1958a. "Report on a Proposed International Standard for a Common Algebraic Language for Digital Computers." In *Computer Programming and Artificial Intelligence,* edited by J. W. Carr, 375–401. University of Michigan Summer School. http://www.softwarepreservation.org/ projects/ALGOL/report/Perlis_Samelson-Proposed_International_Standard -1958.pdf.

———. 1958b. "Preliminary Report: International Algebraic Language." *Communications of the ACM* 1 (12): 8–22. https://doi.org/10.1145/377924.594925.

———. 1959. "Report on the Algorithmic Language ALGOL by the ACM Committee on Programming Languages and the GAMM Committee on Programming." *Numerische Mathematik* 1 (1): 41–60. https://doi.org/10.1007/BF01386372.

Perrier, Alex. 2015. "Jupyter, Zeppelin, Beaker: The Rise of the Notebooks." *Open Data Science.* https://www.opendatascience.com/blog/jupyter-zeppelin-beaker-the-rise-of-the-notebooks/https://www.opendatascience.com/blog/jupyter-zeppelin-beaker-the-rise-of-the-notebooks/.

Pflüger, Jörg. 2002. "Language in Computing." In *Experimenting in Tongues: Studies in Science and Language,* edited by Matthias Dörries, 125–64. Stanford University Press.

"A Plone API." n.d. *Plone Documentation.* https://4.docs.plone.org/external/plone.api/docs/index.html.

Pollock, Sheldon. 2009. "Future Philology? The Fate of a Soft Science in a Hard World." *Critical Inquiry* 35 (4): 931–61. https://doi.org/10.1086/599594.

Pound, Louise. 1924. "Notes on the Vernacular." *American Mercury,* October, 233–37. http://www.unz.org/Pub/AmMercury-1924oct-00233.

"Preliminary Specifications: Programmed Data Processor Model Three (PDP-3)." 1960. Digital Equipment Corporation. http://www.gutenberg.org/ebooks/29461.

Priestley, Mark. 2010. *A Science of Operations: Machines, Logic and the Invention of Programming.* Springer.

Punday, Daniel. 2015. *Computing as Writing.* University of Minnesota Press.

Pym, Anthony. 2014. *Exploring Translation Theories.* 2nd ed. Routledge.

"The Python Language Reference." 2019. Python Software Foundation.

Quittner, Joshua, and Michelle Slatalla. 1998. *Speeding the Net: The Inside Story of Netscape and How It Challenged Microsoft.* Atlantic Monthly Press.

Raabe, David M. 1985. "Strategies in Scanning Robert Frost's Poetry with SNOBOL." In *ICEBOL 85 Proceedings: The Proceedings of the 1985 International Conference on English Language and Literature Applications of SNOBOL and SPITBOL,* 132–43. Dakota State College.

Raley, Rita. 2002. "Interferences: [Net.Writing] and the Practice of Codework." *EBR: Electronic Book Review,* September.

Raskin, Jeffrey F. 1971. "Programming Languages for the Humanities." *Computers and the Humanities* 5 (3): 155–58. https://www.jstor.org/stable/30199400.

Raymond, Eric S. 2004a. "Metasyntactic Variable." *The Jargon File.* http://www.catb.org/jargon/html/M/metasyntactic-variable.html.

———. 2004b. "The Jargon File, Version 4.4.8." *Eric S. Raymond's Home Page.* http://www.catb.org/~esr/jargon/.

Reid, Robert. 1997. *Architects of the Web: 1,000 Days That Built the Future of Business.* John Wiley & Sons.

Reisner, Alex. 2012. "Eff You Foo Bar." *Alex Reisner.* http://www.alexreisner.com/code/eff-you-foo-bar.

Reynolds, J., and J. Postel. 1987. "The Request for Comments Reference Guide." Request for {Comments} ({RFC}) 1000. Network Working Group. https://www.ietf.org/rfc/rfc1000.txt.

Rifkin, Jeremy. 1996. *The End of Work: The Decline of the Global Labor Force and the Dawn of the Post-Market Era.* Putnam.

Ritchie, Dennis M. 1996. "The Development of the C Programming Language." In *History of Programming Languages II*, edited by Thomas J. Bergin and Richard G. Gibson, 671–87. ACM Press, Addison-Wesley.

Robison, Hank, and Alvin Chang. 2013. "America's Incredible Shrinking Information Sector." *Harvard Business Review*, November. http://hbr.org/2013/11/americas-incredible-shrinking-information-sector/.

Rosenberg, Scott. 2008. *Dreaming in Code: Two Dozen Programmers, Three Years, 4,732 Bugs, and One Quest for Transcendent Software.* Three Rivers Press.

Russell, Andrew L. 2014. *Open Standards and the Digital Age: History, Ideology, and Networks.* Cambridge University Press.

Russell, Andrew L., and Lee Vinsel. 2018. "After Innovation, Turn to Maintenance." *Technology and Culture* 59 (1): 1–25. https://doi.org/10.1353/tech.2018.0004.

Ryan, James. 2017. "Grimes' Fairy Tales: A 1960s Story Generator." In *Interactive Storytelling*, edited by Nuno Nunes, Ian Oakley, and Valentina Nisi, 10690:89–103. Springer International Publishing. https://doi.org/10.1007/978-3-319-71027-3_8.

Sachs, Jon. 1976. "Some Comments on Comments." *ACM SIGDOC Asterisk Journal of Computer Documentation* 3 (7): 7–14. https://doi.org/10.1145/1110798.11 10799.

Sack, Warren. 2019. *The Software Arts.* MIT Press.

Said, Edward W. 2004. "The Return to Philology." In *Humanism and Democratic Criticism*, 57–84. Columbia University Press.

Sammet, Jean E. 1969. *Programming Languages: History and Fundamentals.* Prentice-Hall.

Samson, Peter R. 1959. "Foo." *Tech Model Railroad Club Dictionary.* http://www.gricer.com/tmrc/dictionary1959.html.

Samson, Peter. 1966. "PDP-6 LISP." http://hdl.handle.net/1721.1/5899.

Sanders, Ruth H. 1985. "PILOT, SNOBOL and Logo as Computing Tools for Foreign-Language Instruction." *Calico Journal* 3 (2): 41–47. http://www.jstor.org/stable/24156908.

Schildt, Herbert. 2017. *Java: The Complete Reference.* 10th ed. McGraw-Hill Education.

Schiller, Dan. 2014. *Digital Depression: Information Technology and Economic Crisis.* University of Illinois Press.

Schmitt, William F. 1988. "The UNIVAC SHORT CODE." *IEEE Annals of the History of Computing* 10 (1): 7–18. https://doi.org/10.1109/MAHC.1988.10004.

Scholz, Trebor, ed. 2013. *Digital Labor: The Internet as Playground and Factory.* Routledge.

———. 2017. *Uberworked and Underpaid: How Workers Are Disrupting the Digital Economy.* Polity Press.

Schwartz, Oscar. 2019. "Untold History of AI: Invisible Women Programmed America's First Electronic Computer." *IEEE Spectrum*, March. https://spec

trum.ieee.org/untold-history-of-ai-invisible-woman-programmed-americas
-first-electronic-computer.

"Science: Foo-Fighter." 1945. *Time* 45 (3): 72.

Sebesta, Robert W. 2010. *Concepts of Programming Languages*. 9th ed. Addison-
Wesley.

Shannon, Clay. 2003. "Interview with Alan C. Moore." *Embarcadero Developer
Network*. http://edn.embarcadero.com/article/30102.

Shetterly, Margot Lee. 2016. *Hidden Figures: The American Dream and the
Untold Story of the Black Women Mathematicians Who Helped Win the Space
Race*. William Morrow.

Simondon, Gilbert. 2017. *On the Mode of Existence of Technical Objects*. Translated by Cécile Malaspina and John Rogove. University of Minnesota Press.

Simpson, Kyle. 2015. *Up & Going*. 1st ed. O'Reilly Media.

Smith, Andrew. 2018. "Code to Joy." *The Economist*, July. https://www
.1843magazine.com/features/code-to-joy.

Snyder, Bill. 2007. "Java Is Becoming the New Cobol." *InfoWorld*, December.
http://www.infoworld.com/article/2650254/application-development/java-is
-becoming-the-new-cobol.html.

Somaiya, Ravi. 2015. "Josh Tyrangiel Leaving as Editor of Bloomberg Businessweek." *New York Times*, October. https://www.nytimes.com/2015/10/02/
business/media/josh-tyrangiel-leaving-as-editor-of-bloomberg-businessweek
.html.

Srnicek, Nick. 2017. *Platform Capitalism*. Polity.

Srnicek, Nick, and Alex Williams. 2016. *Inventing the Future: Postcapitalism and
a World Without Work*. Rev. and updated ed. Verso.

"Standard ECMA-262: ECMAScript® 2019 Language Specification, 10th Edition." 2019. ECMA International.

Steele, Guy L., Jr. 2022. "Foreword." In *Structure and Interpretation of Computer
Programs, JavaScript Edition*, xiii–xv. MIT Press.

Stepney, Susan. n.d. "How to Shoot Yourself in the Foot." *Susan Stepney*. http://
www-users.cs.york.ac.uk/susan/joke/foot.htm.

Strange, William C. 1985. "Three Computable Poems (And Where They Take
Us)." In *ICEBOL 85 Proceedings: The Proceedings of the 1985 International
Conference on English Language and Literature Applications of SNOBOL and
SPITBOL*, 197–214. Dakota State College.

Stroustrup, Bjarne. 2016. "Bjarne Stroustrup's FAQ." *Bjarne Stroustrup's Homep-
age*. http://www.stroustrup.com/bs_faq.html.

Strunk, William, Jr.1920. *The Elements of Style*. Harcourt, Brace.

Strunk, William, Jr., and E. B. White. 2009. *The Elements of Style*. 4th ed. Long-
man.

Surowiecki, James. 2014. "How Mozilla Lost Its C.E.O." *New Yorker*, April. http://
www.newyorker.com/business/currency/how-mozilla-lost-its-c-e-o.

Swafford, Annie. 2015. "Problems with the Syuzhet Package." *Anglophile in Aca-
demia*. https://annieswafford.wordpress.com/2015/03/02/syuzhet.

Sweeney, Joseph. 2015. "The Use & Abuse of the Word Digital." *Linkedin.* https://www.linkedin.com/pulse/use-abuse-word-digital-dr-joe-sweeney.

Swisher, Kara. 2018. "The Expensive Education of Mark Zuckerberg and Silicon Valley." *New York Times,* August.

tali713. 2009. "Wabbit.scm." https://github.com/tali713/mit-scheme/blob/master/src/wabbit/wabbit.scm.

Tangermann, Victor. 2023. "Elon Musk Just Got Roasted to His Face and Seemed to Have No Idea." *The Byte,* April.

"Temperance Poll Shows 87 Percent of Voters Imbibe." 1938. *The Tech,* December, 1, 4. http://tech.mit.edu/V58/PDF/V58-N48.pdf.

Templeton, Brad, ed. 1995. *The Internet Joke Book.* Peer-to-Peer Communications.

Terranova, Tiziana. 2004. *Network Culture: Politics for the Information Age.* Pluto Press.

Tierney, Matt. 2019. *Dismantlings: Words against Machines in the American Long Seventies.* Cornell University Press.

Tim. 2011. "What Does 'Foo' Mean?" *Stack Exchange: English Language and Usage.* http://english.stackexchange.com/questions/27843/what-does-foo-mean/.

Tosh, Wayne. 1985. "Some SNOBOL Applications in Language and Literature." In *ICEBOL 85 Proceedings: The Proceedings of the 1985 International Conference on English Language and Literature Applications of SNOBOL and SPITBOL,* 82–120. Dakota State College.

"Transclusion: Fixing Electronic Literature." 2007. http://www.youtube.com/watch?v=ohiKTVVtDJA.

Tufekci, Zeynep. 2018. "How Social Media Took Us from Tahrir Square to Donald Trump." *MIT Technology Review,* August.

Uchitelle, Louis, and N. R. Kleinfeld. 1996. "The Price of Jobs Lost." In *The Downsizing of America,* 3–36. 1st ed. Times Books.

Ullman, Ellen. 2012. *Close to the Machine: Technophilia and Its Discontents.* Picador/Farrar, Straus & Giroux.

"Untitled." 1961. *ALGOL Bulletin* 13 (1). http://dl.acm.org/citation.cfm?id=1060935.

US Bureau of Labor Statistics. 2014. "Occupational Outlook Handbook, 2014-15 Edition, Computer Programmers." US Department of Labor. https://web.archive.org/web/20151122074349/http://www.bls.gov/ooh/computer-and-information-technology/computer-programmers.htm.

——. 2015. "Occupational Outlook Handbook, 2016-17 Edition, Computer Programmers." US Department of Labor. http://www.bls.gov/ooh/computer-and-information-technology/computer-programmers.htm.

——. 2020. "Occupational Outlook Handbook, Computer Programmers." US Department of Labor. https://www.bls.gov/ooh/computer-and-information-technology/computer-programmers.htm.

——. 2022. "Occupational Outlook Handbook, Computer Programmers." US Department of Labor. https://www.bls.gov/ooh/computer-and-information-technology/computer-programmers.htm.

Van De Vanter, Michael L. 2002. "The Documentary Structure of Source Code." *Information and Software Technology* 44 (13): 767–82. https://doi.org/10.1016 /S0950-5849(02)00103-9.

Van Wijngaarden, A. 1969. *Report on the Algorithmic Language ALGOL 68*. Mathematisch Centrum.

Van Wijngaarden, A., B. J. Mailloux, J. Peck, and C. H. A. Koster. 1968. "Draft Report on the Algorithmic Language ALGOL 68." *ALGOL Bulletin* 26 (March): 1–84. http://dl.acm.org/citation.cfm?id=1064073.

Verburg, Pieter A. 1998. *Language and Its Functions: A Historico-Critical Study of Views concerning the Functions of Language from the Pre-Humanistic Philology of Orleans to the Rationalistic Philology of Bopp*. Translated by Paul Salmon. J. Benjamins.

Ward, Robert, and Martin Smith. 1998. "JavaScript as a First Programming Language for Multimedia Students." *ACM SIGCSE Bulletin* 30 (3): 249–53. https:// doi.org/10.1145/290320.283557.

Warner, John. 2022. "Freaking Out about ChatGPT—Part I." *Inside Higher Ed*, December.

Warren, Tom. 2016. "Oracle's Finally Killing Its Terrible Java Browser Plugin." *The Verge*, January. http://www.theverge.com/2016/1/28/10858250/oracle -java-plugin-deprecation-jdk-9.

Watson, Ian. 2005. "Cognitive Design: Creating the Sets of Categories and Labels That Structure Our Shared Experience." Rutgers University, Dept. of Sociology. http://www.ianwatson.org/cognitive_design_a4.pdf.

Weaver, Warren. 1949. "Translation." http://www.mt-archive.info/Weaver-1949 .pdf.

———. 1955. "Translation." In *Machine Translation of Languages: Fourteen Essays*, edited by William N. Locke and A. Donald Booth, 15–23. MIT Press.

Wegner, Peter. 1985. "Programming Languages—The First 25 Years." In *Programming Languages: A Grand Tour*, edited by Ellis Horowitz, 4–22. Computer Science Press.

Wegstein, Joseph H. 1956. "Automatic Coding Principles." In *Symposium on Advanced Programming Methods for Digital Computers: Washington, D.C., June 28, 29, 1956*, 3–6. Office of Naval Research, Dept. of the Navy.

Weinberg, Gerald M. 1998. *The Psychology of Computer Programming*. Silver Anniversary Edition. Dorset House.

Weiss, Aaron. 1999. "JavaScripting into the Next Millenniun." *netWorker* 3 (4): 34–35. https://doi.org/10.1145/323409.328683.

"Welcome to the Canvas LMS API Documentation." n.d. *Canvas*. https://canvas .instructure.com/doc/api/index.html.

Wentworth, Harold, and Stuart Berg Flexner, eds. 1960. *Dictionary of American Slang*. Thomas Y. Crowell.

———. 1975. *Dictionary of American Slang*. 2nd supplemented ed. Thomas Y. Crowell.

Wexelblat, Richard L., ed. 1981. *History of Programming Languages*. Academic Press.

"What Does 'foo' Mean?" *Stack Exchange: English Language and Usage,* 2011. https://english.stackexchange.com/questions/27843/what-does-foo-mean/.

Wildes, Karl L., and Nilo A. Lindgren. 1985. *A Century of Electrical Engineering and Computer Science at Mit, 1882-1982.* MIT Press.

Williams, Alex, and Nick Srnicek. 2013. "#ACCELERATE MANIFESTO for an Accelerationist Politics." *Critical Legal Thinking.* http://criticallegalthinking .com/2013/05/14/accelerate-manifesto-for-an-accelerationist-politics/.

Willis, John. 2012. "The Convergence of DevOps." *IT Revolution Press.*

Ynda-Hummel, Ian. 2013. "What Is the Longest Method Name You've Ever Seen?" *Quora.* https://www.quora.com/What-is-the-longest-method-name-youve-ev er-seen.

Yngve, Victor H. 1957. "A Framework for Syntactic Translation." *Mechanical Translation* 4 (3): 59–65.

——. 1958. "A Programming Language for Mechanical Translation." *Mechanical Translation* 5 (1): 25–41.

——. 1962. "COMIT as an IR Language." *Communications of the ACM* 5 (1): 19–28. https://doi.org/10.1145/366243.366720.

——. 1963. "COMIT." *Communications of the ACM* 6 (3): 83–84. https://doi.org/10 .1145/366274.366291.

INDEX

1995, as historically significant year, 50, 134, 136, 140–143, 171
2016, as historically significant year, 2, 49–50, 132–133, 151

Abbate, Janet, 8, 47–48, 77
Ada (programming language), 83
Adams, Charles W., 32, 35–37, 41
Advanced Research Projects Agency (ARPA), 76
Agile software development, 159–160
Algol (programming language), 7, 57–60, 66–68, 83, 87, 101, 109, 118, 123, 128, 138, 184n14, 189n11, 190n20, 191n20, 192n3
"ALGOL 68 with Fewer Tears" (Lindsey), 66–68
algorithms, 17, 19, 46, 87, 90, 192n5
API (Application Programming Interface), 22, 81, 88–90
APL (programming language), 103, 123, 128
array (data structure), 105–106, 130

assembly, 37, 40, 57, 59–60, 93, 115, 122, 128–129, 131, 139, 189n10, 195n8. *See also operation codes*
autology, 4–5, 18, 180. *See also* heterology
automatic coding, 35–36, 45, 47, 49
automatic programming. *See* automatic coding
automation: and productivity, 178; and unemployment, 163–165; as a myth, 20, 34, 167; as incomplete, 23, 166–167; classical concept of, 34; computer science and, 5, 10; discourse and theory, 165–166; maintenance of, 23; programming's special role in, 5, 13–14, 20 22, 33–34, 37, 65, 166; recursive, 5, 45–46, 129, 157, 160, 167–168, 180, 183n6; semi-, 29; shitty, 168; software deployment, 160; spotty, 168; usage history of word, 164. *See also* programming; programming languages

object (data structure), 106, 130, 140

Objective-C (programming language), 83, 95, 150, 189n12

OCaml (programming language), 149

offshoring. *See* outsourcing

OpenAI Inc., 18–19, 22, 65. *See also* ChatGPT

operation codes, 5, 35, 37, 39–40, 43, 51, 53, 57, 59, 84, 92–93, 129, 138, 139, 195n8. *See also* assembly

outsourcing, 49, 171, 175

Pascal (programming language), 69–70, 83, 128

pattern matching (programming language feature), 116–118, 122, 127

Perl (programming language), 58, 83, 140, 149

philology: and science and technology studies (STS), 7–9; as distinct from historiography, 14–15, as distinct from philosophy, 14; as humanist tradition, 7; etymology of, 14; in the history of computing, 35; main characteristics of, 13; of programming languages, 15–16, 18; scope of, 13–16

philosophy, 6, 14–16

PHP (programming language), 83, 89–90, 189n12

PL/I (programming language), 60, 103, 123

platform studies, 5

pointers, 94–96

postcritique, 17

Priestley, Mark, 7

program code. *See* source code

program execution, 53, 56, 64–65, 127

"Programmer as Reader" (Goldberg), 63–64

programming: academic, 59, 90, 107–110, 119; as the automation of automation, 5, 20–21; as literary composition, 68–70; as multilingual, 4; as translational, 4; earliest history of, 168–169; essayistic, 68–70; gender and, 30, 47–48, 168; in real time, 64–65; instruction, 59, 64, 82, 90–91, 97, 133, 136–137; literate, 10, 63, 65–72, 74, 190n23; masculinization of, 47–48; nonuniformity, 62; novice and non-professional, 90, 94, 96–99, 100, 103, 107–109, 119, 124–125, 127, 134, 138, 140, 145, 147; object-oriented paradigm, 64, 89, 94, 134–135, 191n14; professional, 134, 136, 138, 151–152; psychological "set" in, 61–62; structured, 59, 68, 101, 103, 189n13, 189n15; style, 59–61, 63, 68; systems, 95, 122, 134. *See also* software

programming languages: and foreign languages, 11; as codes, 12, 17; as focus of this book, 13–14; Backus-Naur form of notation for syntax of, 86, 109, 192n3, 193n9; cultures of, 5–6; design of, 64, 80, 93, 101, 104; domain-specific, 115, 123–124, 172; functional, 83, 145–146; general-purpose, 108–109, 117, 122–124, 148, 150; hierarchy of, 5, 34, 92–93, 129, 192n28, 195n8; higher-level, 5, 34–35, 51, 56–57, 59, 78, 92–93, 103–104, 106, 129–131, 138, 195n8; histories and historiographies of, 7–8, 14, 38, 63–64, 98, 101, 130–131; imperative, 83, 110, 116, 145–146; implementation of, 122, 130; learning of, 11–12; logic, 83; names of, 113–114; object-oriented, 83, 145–146; reserved words in, 54, 80, 91, 93; scripting, 137–140, 143, 147, 149; specifications, 191n12; string and

The authorized representative in the EU for product safety and compliance is:
Mare Nostrum Group
B.V Doelen 72
4831 GR Breda
The Netherlands

www.ingramcontent.com/pod-product-compliance
Lightning Source LLC
Chambersburg PA
CBHW020858270326
41928CB00006B/764